Lecture Notes in Computer Science   13460

More information about this subseries at https://link.springer.com/bookseries/8183

Marina L. Gavrilova · C. J. Kenneth Tan (Eds.)

Transactions on Computational Science XXXIX

Springer

Editors-in-Chief
Marina L. Gavrilova
University of Calgary
Calgary, AB, Canada

C. J. Kenneth Tan
Sardina Systems OÜ
Tallinn, Estonia

ISSN 0302-9743   ISSN 1611-3349   (electronic)
Lecture Notes in Computer Science
ISSN 1866-4733   ISSN 1866-4741   (electronic)
Transactions on Computational Science
ISBN 978-3-662-66490-2   ISBN 978-3-662-66491-9   (eBook)
https://doi.org/10.1007/978-3-662-66491-9

This Springer imprint is published by the registered company Springer-Verlag GmbH, DE, part of Springer Nature
The registered company address is: Heidelberger Platz 3, 14197 Berlin, Germany

LNCS Transactions on Computational Science

Computational science, an emerging and increasingly vital field, is now widely recognized as an integral part of scientific and technical investigations, affecting researchers and practitioners in areas ranging from aerospace and automotive research to biochemistry, electronics, geosciences, mathematics, and physics. Computer systems research and the exploitation of applied research naturally complement each other. The increased complexity of many challenges in computational science demands the use of supercomputing, parallel processing, sophisticated algorithms, and advanced system software and architecture. It is therefore invaluable to have input by systems research experts in applied computational science research.

Transactions on Computational Science focuses on original high-quality research in the realm of computational science in parallel and distributed environments, also encompassing the underlying theoretical foundations and the applications of large-scale computation.

The journal offers practitioners and researchers the opportunity to share computational techniques and solutions in this area, to identify new issues, and to shape future directions for research, and it enables industrial users to apply leading-edge, large-scale, high-performance computational methods.

In addition to addressing various research and application issues, the journal aims to present material that is validated – crucial to the application and advancement of the research conducted in academic and industrial settings. In this spirit, the journal focuses on publications that present results and computational techniques that are verifiable.

Scope

The scope of the journal includes, but is not limited to, the following computational methods and applications:

- Aeronautics and Aerospace
- Astrophysics
- Big Data Analytics
- Bioinformatics
- Biometric Technologies
- Climate and Weather Modeling
- Communication and Data Networks
- Compilers and Operating Systems
- Computer Graphics
- Computational Biology
- Computational Chemistry
- Computational Finance and Econometrics
- Computational Fluid Dynamics

- Computational Geometry
- Computational Number Theory
- Data Representation and Storage
- Data Mining and Data Warehousing
- Information and Online Security
- Grid Computing
- Hardware/Software Co-design
- High-Performance Computing
- Image and Video Processing
- Information Systems
- Information Retrieval
- Modeling and Simulations
- Mobile Computing
- Numerical and Scientific Computing
- Parallel and Distributed Computing
- Robotics and Navigation
- Supercomputing
- System-on-Chip Design and Engineering
- Virtual Reality and Cyberworlds
- Visualization

Editorial

The Transactions on Computational Science journal is published as part of the Springer series, Lecture Notes in Computer Science, and is devoted to a range of computational science issues, from theoretical aspects to application-dependent studies and the validation of emerging technologies.

The journal focuses on original, high-quality research in the realm of computational science in parallel and distributed environments, encompassing the theoretical foundations and the applications of large-scale computations and massive data processing. Practitioners and researchers share computational techniques and solutions, identify new issues that shape the future directions for research, and enable industrial users to apply the presented techniques.

The current journal issue is devoted to research on geometric modeling, visual object detection, cloud service utilization, pattern recognition, processing arrays, and image classification using bio-heuristic optimization. It is comprised of five papers submitted as regular papers to the journal.

The first article of the issue, "Degradable Self-Restructuring of Processor Arrays by Direct Spare Replacement", introduces an efficient way to create a processor array by direct spare replacement. With the increase in the number of processing elements in modern highly integrated parallel systems, there has been a growing focus on designing an efficient self-restructuring method to tolerate faulty processing elements. The authors tackle this problem by presenting a self-restructuring method for mesh-connected processor arrays with spares on the orthogonal sides, based on combining the redundancy and degradation methodologies.

The second article "Structural Composite Feature Triangulation for Visual Object Search" delivers a new approach to visual object search. Its key contributions center on the proposed Structural Composite Feature Triangulation method. The method mines and detects composite structures from Delaunay triangulation, in order to subsequently perform detection and localization of multiple object instances within an image. Experimental results on two datasets convincingly demonstrate that the proposed method can detect multiple object instances and is able to effectively retrieve the relevant images.

The third article "Study of Malaysian Cloud Industry and Conjoint Analysis of Healthcare and Education Cloud Service Utilization" addresses the Malaysian infrastructure and capability of adopting cloud computing services. It also provides statistical analysis and insights into the country's internet infrastructure and internet uses, fixed and mobile broadband deployment, and usage of international bandwidth for improvising cloud services. In addition, the paper discusses pertinent privacy and security laws, as relevant to the adoption of cloud services across regions and borders.

The fourth article "Algorithms for Generating Strongly Chordal Graphs" presents an elegant solution to an open problem of weakly chordal graphs from arbitrary graph generation. The authors propose a method to reduce graph G to a chordal graph H by adding fill-edges and using the minimum vertex degree heuristic. Next, the authors devise an algorithm for deleting edges from a weakly chordal graph that preserves weak chordality.

The fifth article "A Novel Machine Learning Framework for Covid-19 Image Classification with Bioheuristic Optimization" presents a novel bio-inspired algorithm for image classification. The authors introduce a new framework for COVID-19 image categorization that utilizes deep learning and bio-inspired optimization techniques. The experimental results indicate that the proposed approach improves performance in terms of classification accuracy while also achieving a significant reduction in computational costs.

The sixth article "An Unsupervised DNN Embedding System for Image Clustering" presents a new concept of an unsupervised AutoEmbedder – UautoEmbedder, which is trained in an unsupervised fashion. Through rigorous research, the authors establish that UAutoEmbedder is an ideal architecture for data augmentation dependent unsupervised learning. The method performance is validated on image datasets.

We thank all reviewers for their diligence in making recommendations and evaluating revised versions of the papers presented in this journal issue. We would also like to thank all of the authors for submitting their papers to the journal and the associate editors for their valuable work.

It is our hope that this collection of five articles presented in this special issue will be a valuable resource for Transactions on Computational Science readers and will stimulate further research in the vibrant area of computational science theory and applications.

July 2022

Marina L. Gavrilova
C. J. Kenneth Tan

LNCS Transactions on Computational Science –
Editorial Board

Contents

Degradable Self-restructuring of Processor Arrays by Direct Spare Replacement

Itsuo Takanami[1] and Masaru Fukushi[2]($\boxtimes$)

[1] Department of Technology, Yamaguchi University,
Tokiwadai 2-16-1, Ube 755-0097, Japan
iftakanami@comet.ocn.ne.jp
[2] Graduate School of Sciences and Technology for Innovation,
Yamaguchi University, Tokiwadai 2-16-1, Ube 755-0097, Japan
mfukushi@yamaguchi-u.ac.jp

Abstract. With the increase in the number of processing elements
(PEs) in modern highly integrated parallel systems, there has been a
growing importance for designing an efficient self-restructuring method
to automatically tolerate faulty PEs. In this paper, we present a self-
restructuring method for mesh-connected processor arrays with spares
on the orthogonal sides, based on the cooperation of the redundancy
and degradation approaches. The redundancy approach replaces faulty
PEs with spares, while the degradation approach deletes rows and/or
columns of the arrays. First, we formalize the spare assignment problem in the redundancy approach as a matching problem in graph theory.
Then, if no matching, i.e., no valid spare assignment, is found for an
array with faulty PEs, rows and/or columns are deleted from the array
so that a matching is successfully found for a degraded subarray. Finally,
hardware circuits to realize the above process are presented. This leads
to the realization of degradable self-restructuring of processor arrays,
and implies that the proposed method is useful in enhancing especially
the run-time reliability and availability of processor arrays in mission
critical applications where first self-reconfiguration is required without
an external host computer.

Keywords: Fault-tolerance · Mesh array · Direct spare replacement ·
Self-restructuring · Built-in circuit

1 Introduction

Recently, high-speed and high-quality technologies for processing many kinds of
information have become essential. It is expected that higher-speed and higher-
quality technologies will become more and more necessary in the future. For these
needs, how to realize massively parallel computing systems has been studied in
the literature. A mesh-connected processor array (PA) is a kind of form of massively parallel computing systems. Mesh-connected PAs consisting of processing

© Springer-Verlag GmbH Germany, part of Springer Nature 2022
M. L. Gavrilova and C. J. K. Tan (Eds.): Trans. on Comput. Sci. XXXIX, LNCS 13460, pp. 1–21, 2022.
https://doi.org/10.1007/978-3-662-66491-9_1

elements (PEs) have regular and modular structures which are very suitable for most signal and image processing algorithms.

In such a situation, as VLSI technology has developed, the realization of parallel computing systems using multi-chip module (MCM) e.g., [1], wafer scale integration (WSI) e.g., [2] or network-on-chip (NoC) e.g., [3] has been considered so as to enhance the performance of the computing systems and to decrease energy consumption and sizes, and so on. In such realization, entire or significant parts of PEs and inter-connections among them are implemented on a VLSI chip or wafer. Therefore, the yield and/or reliability of the system may become drastically low if there is no strategy for coping with defects and faults. To restore the correct computation capabilities of PAs, it must be restructured appropriately so that the defective PEs are eliminated from the computation paths, and the working PEs maintain correct logical connectivity among them.

Various strategies to restructure a faulty physical PA into a fault-free target logical PA are described in the literature. They have mainly focused on two approaches of the redundancy, e.g., [4–9] and the degradation, e.g., [10–15].

In the redundancy approach, some spare PEs are incorporated to replace faulty PEs so that the original size of the array is preserved. However, if all faulty PEs in an array are not replaced by spares due to the insufficient number of spares and/or the restructuring ability, the array is considered to be faulty and not used even if there exists a lot of healthy PEs.

In the degradation approach, all PEs are considered in a uniform manner in the absence of spare PEs to construct a subarray, avoiding faulty PEs but keeping the mesh structure so that it's size is as large as possible. Then two-dimensional degradable arrays linked with four-port switches have been extensively studied. Kuo and Chen studied the degradable reconfiguration problem under the certain routing constraints and showed that most reconfiguration problems under the rerouting constraints are NP-complete [16]. Low et al. proposed an optimal algorithm, termed GCR to construct a maximum logical array (MLA) containing the selected rows [17]. Fukushi et al. used genetic approaches to evolve rerouting strategies for constructing logical rows/columns in designing the MLA [10,11].

These two approaches have been studied separately. So, it is expected that a PA which is judged to be irreparable by the redundancy approach can be reused by incorporating the degradation approach, though it may lead to reduce their sizes.

In this paper, we present a restructuring method to apply the redundancy and the degradation approaches together to the direct spare replacement scheme (DSRS) [18]. This method can be realized in hardware for the fast reconfiguration of PAs without control by an external host computer. As far as we know, this is the first work which enables degradable self-reconstruction with the combination of two different approaches.

In Sect. 2, we present the DSRS. In Sect. 3, we formalize the strategies for deciding that by which healthy spares faulty PEs should be replaced as a matching problem in graph theory, and present an algorithm for restructuring the arrays with faulty PEs in a convenient form for finding subarrays without unrepaired faulty PEs by degradation. In Sect. 4, the measures to evaluate the proposed method are presented. In Sect. 5, hardware circuits to realize the proposed method are presented. Section 6 is a conclusion.

2   DSRS and Repairability Condition

Figure 1 shows the DSRS to be treated here where linear arrays of spares are arranged on the 0-th row and 0-th column of an array with a size of 4×4, respectively. If PE p_{ij} at the i-th row and j-th column (the notation $p(i, j)$ is also used) is faulty, it is directly replaced either by spare p_{0j} on the 0-th row or by spare p_{i0} on the 0-th column. Note that if a spare PE is faulty, it is considered to be replaced by itself.

$\boxed{\text{S}}$: spare PE

Fig. 1. DSRS

Notation:

- If a faulty PE is replaced by a spare PE, it is said to be **repaired**, otherwise **unrepaired**.
- For a set of faulty PEs which is often called a **fault pattern**, if all the faults in the set can be repaired at the same time, the fault pattern is said to be **repairable**, otherwise **irreparable**.

For the DSRS, we formalize a strategy for deciding that by which nonfaulty spares faulty PEs should be replaced as a matching problem in graph theory.

For the array in Fig. 1, we construct the following bipartite graph G called a compensation graph.

Let $V_f = \{p'_{ij} | 0 \le i \le N, 0 \le j \le N, 1 \le i + j,\ p_{ij}\ \text{is faulty}\}$, $V_c = \{p_{10}, ..., p_{N0}\}$ which is a set of spares in the upper side of the array and $V_r = \{p_{01}, ..., p_{0N}\}$ which is in the left side of the array. Then, $G = (V, E)$, where $V = V_f \cup (V_c \cup V_r)$, $E = \{(p'_{ij}, p_{i0}) | 1 \le i \le N, 0 \le j \le N,\ 1 \le i + j,\ p_{ij}\ \text{is faulty}\} \cup \{(p'_{ij}, p_{0j}) | 0 \le i \le N, 1 \le j \le N,\ 1 \le i + j,\ p_{ij}\ \text{is faulty}\}$. V_f is called a (vertex) set of faulty PEs, $(V_c \cup V_r)$ a (vertex) set of spares and E a set of edges implying replacement relation, respectively. Note that a faulty spare PE is considered to be replaced by itself.

It is clear that the following holds.

Lemma 1. The set of faulty PEs V_f is repairable by replacement if and only if a matching from V_f to $(V_c \cup V_r)$ exists. For such a matching M, faulty p_{ij} is replaced by spare p_{i0} if $(p_{ij}, p_{i0}) \in M$, and by spare p_{0j} if $(p_{ij}, p_{0j}) \in M$.   □

Notation:

- Let $G = (V, E)$ be a bipartite graph where $V = V_1 \cup V_2$ and $V_1 \cap V_2 = \phi$ For $S(\subseteq V_1)$, let $\psi(S) = \{v \in V_2 | (w, v) \in E, w \in S\}$. The degree of a vertex u which is the number of edges incident to u is denoted as $deg(u)$.

It is seen that the degree of any faulty vertex in a compensation graph is equal to or less than two. Note that the degree of a nonspare faulty vertex is two and that of a spare faulty vertex is one. Using the fact, the repairability condition is given as follows [18].

Theorem 1 *(Repairability theorem).* Let $G = (V, E)$ be a bipartite graph such that $V = V_1 \cup V_2$, $V_1 \cap V_2 = \phi$ and $E \subseteq V_1 \times V_2$ where the degree of any vertex in V_1 is equal to or less than two. We partition the maximal subgraph of G with the vertex set $V_1 \cup \psi(V_1)$ into connected components and denote the vertex sets in V_1 of the connected components as $C_1, C_2, ..., C_m$ (for each C_p, $C_p \subseteq V_1$, $\psi(C_p) \subseteq V_2$, and for $i \neq j$, $(C_i \cup \psi(C_i)) \cap (C_j \cup \psi(C_j)) = \phi$). Then, the repairability condition is as follows.

There exists a matching from V_1 to V_2 if and only if $|C_i| \leq |\psi(C_i)|$ for all C_i holds, where $|C|$ means the number of elements in C.   □

According to Theorem 1, we can judge whether a PA is repairable. Theorem 1 is also expressed in a convenient form for restructuring a PA with degradation, which is used in the following DR-ALG.

Lemma 2. For any v of degree 1 in V_2 and $(w, v) \in E$, let G' be the graph obtained by removing $\{w, v\}$ from V and the edges incident to w or v. Then, there exists a matching from V_1 to V_2 in G if and only if there exists a matching from $V_1 - \{w\}$ to $V_2 - \{v\}$ in G'.   □
Proof: see [18].

Lemma 3. Let the degree of any vertex in $\psi(C_i)$ be equal to or greater than two. Then,

1. If there is a vertex in $\psi(C_i)$ whose degree is greater than two, there exists no matching from V_1 to V_2.
2. If the degree of every vertex in $\psi(C_i)$ is two and there exists a vertex of degree 1 in C_i, there exists no matching from V_1 to V_2.
3. If the degree of every vertex in $\psi(C_i)$ is two and there exists no vertex of degree 1 in C_i, there exists a matching from V_1 to V_2.   □

Proof: see [18].

3   Degradable Restructuring Algorithm

Now, the degradation of an array is done by functionally (or logically) deleting rows and/or columns of an array. Then, the degradation changes the compensation graph as follows.

Suppose that the i-th row r_i (the i-th column c_i) corresponding to p_{i0} (p_{0i}) is functionally deleted. Then, p_{i0} (p_{0i}) and faulty vertices p's such that (p', p_{i0})s $((p', p_{0i})$s) are deleted, including the edges incident to p's. Such a deletion of a row (column) may reduce the degrees of some vertices in V_2, which is called **D-deletion** in the following. With the situation in mind, a degradable restructuring algorithm (shortly written as **DR-ALG**) is given as follows.

DR-ALG

Step 1 Let $M = \phi$ (empty set), $E' = E$, $V_1' = V_1$ and $V_2' = V_2$.

Step 2 While there is a vertex v with $deg(v) = 1$ in V_2', do the following.
For $(w, v) \in E'$, let $M = M \cup \{(w, v)\}$, $\hat{E} = \{(w, \hat{v})|(w, \hat{v})$ in $E'\}$, $E' = E' - \hat{E}$, $V_1' = V_1' - \{w\}$ and $V_2' = V_2' - \{v\}$.

Step 3 If $V_1' = \phi$, M is a matching from V_1 to V_2. Then, go to Step 6.

Step 4 If there is a vertex v in V_2' whose degree is more than 2, select and D-delete vertices in V_2' so that the degree of v becomes less than or equal to 2. Further, if there is a vertex u in V_1' whose degree is 1, D-delete w where (u, w) in E and go to Step 2, otherwise go to Step 5.
Note that the degree of any faulty nonspare vertex except faulty spare vertex is two and that of any faulty spare vertex is one. Furthermore, this property does not change in the operation described in Lemma 2.

Step 5 Let $\hat{G} = (\hat{V}_1, \hat{V}_2)$ be a compensation graph to be obtained after D-deleting rows or columns in G. Then, there is a matching in $\hat{G}$ and there is a closed cycle in each derived connected component $\hat{C}_i$, from which just two different matching in $\hat{C}_i$ are derived. Choose one of them which is denoted as M_i. Let $M = M \bigcup\{\cup_i M_i\}$. Then M is a matching from $\hat{V}_1$ to $\hat{V}_2$.

Step 6 The algorithm ends.   □

The following is the detailed process for executing DR-ALG by hardware in mind, where spare PEs are located in the upper and left sides as shown in Fig. 1.

Notation:
- The numbers of unrepaired faulty PEs including a spare in a row r and a column c are denoted as n_f^r and n_f^c, respectively.
(Execution of DR-ALG (shortly written as **EDR-ALG**) for an $X \times Y$ array)

Step I While there is an unrepaired faulty PE in a row or column $*$ with $n_f^* = 1$, do the following (a) and (b) alternatively.

(a) Count the number of unrepaired faulty PEs (including a spare PE) toward a spare in each column. This is done in parallel for all columns. Then, replace a faulty PE in a column c with $n_f^c = 1$ by the spare in the upper side and set n_f^c to 0.

(b) Count the number of unrepaired faulty PEs (including a spare PE) toward a spare in each row. This is done in parallel for all rows. Then, replace a faulty PE in a row r with $n_f^r = 1$ by the spare in the left side and set n_f^r to 0.

Note that a faulty spare is considered to be replaced by itself.

This step corresponds to Step 2 in DR-ALG.

Step II If there is a column or row such that the spare is faulty and $n_f \geq 2$, do the D-deletion process as follows. Otherwise, go to Step III.

This step corresponds to Step 4 in DR-ALG.

D-deletion process
Notation:

- r_i $(1 \leq i \leq X)$ denotes the i-th row.
- c_i $(1 \leq i \leq Y)$ denotes the i-th column.
- α is a variable representing a row or column such as r_j or c_j.
- $n_f^{r_i}$ $(n_f^{c_i})$ is the number of unrepaired faulty PEs including the spare in the i-th row (column).
- $\hat{n}_f^{r_i}$ $(\hat{n}_f^{c_i})$ is the number of faults in the i-th row (column) that should be D-deleted to be repaired. $\hat{n}_f^{r_i}$ and $\hat{n}_f^{c_i}$ are defined as below.

  Let $R_{F(2)}$ be a set consisting of r_i's such that $n_f^{r_i} \geq 2$ and the spare p_{i0} is faulty. For $r_i \in R_{F(2)}$, $\hat{n}_f^{r_i} = n_f^{r_i} - 1$

  Let $C_{F(2)}$ be a set consisting of c_i's such that $n_f^{c_i} \geq 2$ and the spare p_{0i} is faulty. For $c_i \in C_{F(2)}$, $\hat{n}_f^{c_i} = n_f^{c_i} - 1$.

  Let $R_{F(3)}$ be a set consisting of r_i's where $n_f^{r_i} \geq 3$ and the spare p_{i0} is healthy. For $r_i \in R_{F(3)}$, $\hat{n}_f^{r_i} = n_f^{r_i} - 2$.

  Let $C_{F(3)}$ be a set consisting of c_i's where $n_f^{c_i} \geq 3$ and the spare p_{0i} is healthy. For $c_i \in C_{F(3)}$ $\hat{n}_f^{c_i} = n_f^{c_i} - 2$.

  In the following, $R_{F(2)} \cup R_{F(3)}$ and $C_{F(2)} \cup C_{F(3)}$ are simply denoted as R_F and C_F, respectively.

If $R_F \cup C_F = \phi$ (i.e., empty), D-deletion process is ended and go to Step IV.

Step III Choose α in $R_F \cup C_F$ and D-delete the row or column corresponding to α, where how to choose α is proposed below. Then go to Step I.

Three methods to choose α are proposed as follows. Note that in executing the process, one method is chosen, which affects the subsequent degradation process and thus the size of a subarray to be obtained.

- Method 1
  Choose α in $R_F \cup C_F$ arbitrarily. In this method, $R_F \cup C_F$ obtained if the n_f's are suppressed to 3 is the same as that by nonsuppression since a row or column in which n_f is greater than 3 is included in $R_F \cup C_F$. This property is useful in hardware realization to be mentioned in Sect. 4.

Table 1. n_f and $\hat{n}_f$ for Fig. 3

i	1	2	3	4	5	6	7	8
$n_f^{r_i}$	0	3	3	2	2	5	0	0
$\hat{n}_f^{r_i}$	–	1	2	–	–	4	–	–
$n_f^{c_i}$	0	2	2	2	0	4	2	2
$\hat{n}_f^{c_i}$	–	–	1	–	–	2	–	–

Table 2. n_f and $\hat{n}_f$ for Fig. 4

i	1	2	3	4	5	6	7	8
$n_f^{r_i}$	0	2	2	2	0	–	0	0
$\hat{n}_f^{r_i}$	–	–	1	–	–	–	–	–
$n_f^{c_i}$	0	0	2	0	0	2	0	2
$\hat{n}_f^{c_i}$	–	–	1	–	–	–	–	–

- Method 2
  Choose α in $R_F \cup C_F$ such that $\hat{n}_f^{\alpha}$ is maximal.
- Method 3

  (1) For each α in $R_F \cup C_F$, the row or column corresponding to α is tried
  to be D-deleted. Then, let the fault pattern obtained be $P_{D(\alpha)}$.
  (2) Apply Step I to $P_{D(\alpha)}$, compute $R'_F \cup C'_F$ and D-delete the row or
  column corresponding to α such that $|R'_F \cup C'_F|$ is minimal.

Step IV A spare PE by which each unrepaired faulty PE in closed cycles in Step
5 in DR-ALG will be replaced is determined.
Step V The restructuring process with degradation is ended.

To help understanding EDR-ALG, Fig. 2 is an instance of an 8×8 array with
21 faulty PEs where $\times$'s denote faulty PEs and the arrows are the directions of
the compensation for them after Step I in EDR-ALG has been executed. As
the result, we have the fault pattern as in Fig. 3 in which the black squares are
ones which have replaced the faulty PEs shaded with gray. Then, the number
of unrepaired faults in each row or column becomes as in Table 1. So, $R_{F(2)} =
\{r_3, r_6\}$, $R_{F(3)} = \{r_2\}$, $C_{F(2)} = \{c_3\}$ and $C_{F(3)} = \{c_6\}$. By Method 1, α should
be chosen arbitrarily. By Method 2, r_6 is chosen as α since $\hat{n}_f^{r_6}$ is maximal. By
Method 3, applying (1) and (2) in Method 3, r_6 is chosen since $|R'_F \cup C'_F|$ is 2 and
minimal, where $|R'_F \cup C'_F|$'s for r_2, r_3, c_3 and c_6 are 3, 3, 4 and 3, respectively,
which can easily be checked.

Now, choosing and D-deleting r_6 in the subarray in Fig. 3, Step I in EDR-
ALG is applied to the subarray obtained where the arrows in Fig. 4 indicate
the directions of compensation for the faulty PEs after Step I in EDR-ALG has
been executed. Then, the number of unrepaired faults remained in each row or

column becomes as in Table 2. Then, either r_3 or c_3 can be D-deleted. Here, suppose D-delete c_3. Then, the subarray as in Fig. 5 is obtained, $R_F \cup C_F$ is empty, and EDR-ALG is ended. Finally, we have an array restructured with degradation whose size is 7×7 where the arrows in Fig. 6 indicate the directions of compensation for the faults and the squares marked with + are PEs D-deleted, which are bypassed vertically and horizontally.

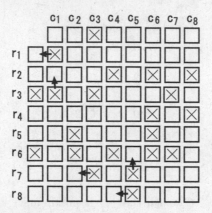

Fig. 2. Arrows are directions of compensation for faults after Step I in EDR-ALG has been executed

We have presented a degradable restructuring algorithm for mesh arrays by direct spare replacement and described how the algorithm is executed.

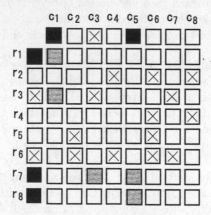

Fig. 3. Black squares are spares which have replaced faulty PEs marked with gray

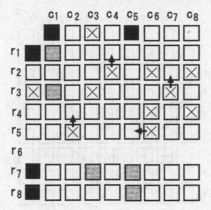

Fig. 4. Subarray obtained by D-deleting the 6th row and arrows are directions of compensation for faults after Step I in EDR-ALG has been executed

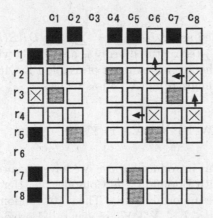

Fig. 5. One of subarrays finally obtained by D-deletion

4 Evaluation of Degradable Restructuring

Several terms and measures to evaluate DR-ALG are introduced as follows.

1. A fault pattern P is called to be $D(r, s)$-restructurable if it is restructured by D-deleting r rows and s columns by EDR-ALG.
2. N_a is the number of all PEs for an array with size of $X \times Y$, i.e., $(X + 1) \cdot (Y + 1) - 1$.
3. $N_e(k)$ is the number of examined fault patterns which have k faulty PEs.
4. $N_{res}(k; r, s)$ is the number of $D(r, s)$-restructurable fault patterns which have k faulty PEs.
5. $N_{res}(k; d)$ is the number of $D(r, s)$-restructurable examined fault patterns which have k faulty PEs and $d = r + s$, i.e., $N_{res}(k; d) = \sum_{r+s=d} N_{res}(k; r, s)$.
6. D-restructured rate $DRR(k; d)$ is defined as $DRR(k; d) = N_{res}(k; d)/N_e(k)$.

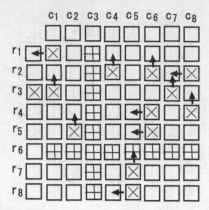

Fig. 6. D-restructured array with usable size of 7×7 where faulty PEs and boxes with mark + are bypassed horizontally and vertically

7. $SDRR(k; d)$ is defined as $SDRR(k; d) = \sum_{0 \leq i < d} DRR(k; i)$.
8. Noting that a $D(r, s)$-restructured array is a subarray with size of $(X - r) \times (Y - s)$, we define the average number of usable PEs $AU(k)$ for the number of faulty PEs k, normalized by $(X \cdot Y)$ as follows.

$$AU(k) = 1/(X \cdot Y) \sum_{0 \leq r,s} (X - r)(Y - s) \cdot N_{res}(k; r, s)/N_e(k).$$

To evaluate the terms introduced above, we have executed Monte Carlo simulations, using a PC with Borland C++ Compiler 5.5. Here, it is assumed that all the PEs may become uniformly faulty. Then, 10^6 random fault patterns each with k faulty PEs for $1 \leq k \leq X + Y$ are generated provided that each PE in an array has the same reliability p.

Figure 9 shows $AU(k)$ for the cases that $X = Y = 8$, 16 and 32. It is seen that $AU(k)$'s increase in the order of M1, M2 and M3. However, for easiness of hardware realization, Method 1 is adopted in the next section because the differences among them are not considered to be large.

5 Hardware Realization

Figure 10 is an example of a PA with switches for replacing faulty PEs. Each PE except spare PEs has four switches around it and if it is faulty, these are switched according to that a spare PE which replaces it is located in the upper or left side where PEs in D-deleted rows or columns are bypassed horizontally and vertically. The procedure EDR-ALG will be realized by hardware circuits for this replacement scheme. This is done by two digital circuits. One is NET-1 and another is NET-2, which are the modified version of [18]. First, NET-1 outputs a signal whether a PA with faulty PEs is repairable without D-deletion, that

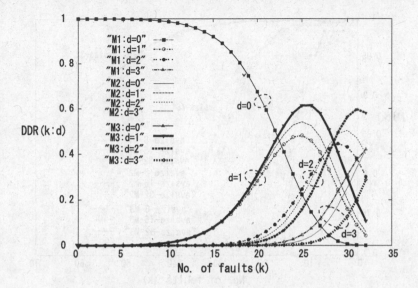

Fig. 7. D-restructured rate for arrays with sized of 16×16

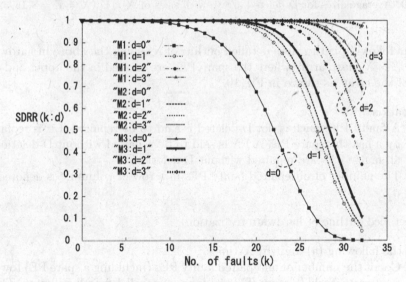

Fig. 8. SDRR for arrays with size of 16×16

is, any faulty PE is replaced by a healthy spare PE. Next, if the array is not repairable without D-deletion, a signal indicating so is output from the terminal "UNRP" and D-deletion process starts. NET-2 determines a direction to a spare by which each faulty PE in the closed cycles obtained in Step IV of EDR-ALG should be replaced (Figs. 11 and 13).

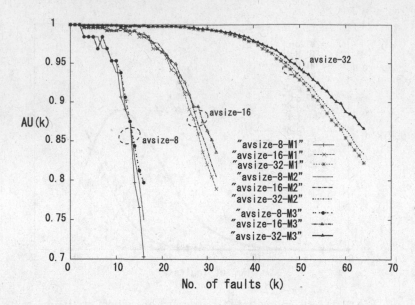

Fig. 9. Average sizes for D-deleted arrays with sizes of $X \times Y$ ($X = Y = 8, 16, 32$)

The following is the more detailed outline of realizing the above in hardware for an $X \times X$ mesh array where the spare PEs are located in the upper and left sides of the array as shown in Fig. 10.

Notation:
- A faulty PE which is not D-deleted is said to be repaired if it is replaced by a healthy spare PE. A PA is said to be repaired without D-deletion if all faulty PEs are repaired without D-deletion.
- The number of unrepaired faulty PEs in a row or column * is denoted as n_f^*

(Detailed outline of hardware realization)

1. Do the following (a) and (b) X times.
   (a) Count the number of unrepaired faulty PEs (including a spare PE) toward a spare in each column. This is done in parallel for all columns. Then, replace a faulty PE in a column with $n_f = 1$ by a spare in the upper side and set n_f to 0.
   (b) Count the number of unrepaired faulty PEs (including a spare PE) toward a spare in each row. This is done in parallel for all rows. Then, replace a faulty PE in a row with $n_f = 1$ by a spare in the left side and set n_f to 0.
   This step corresponds to Step I in EDR-ALG.
2. (i) If there is a column or row with $n_f \geq 3$ or

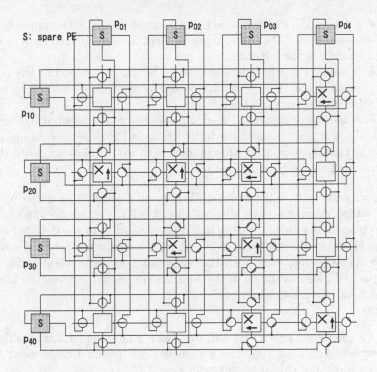

Fig. 10. An example of a mesh array with switches for replacing faulty PEs

(ii) there is a column or row with $n_f = 2$ and the spare in the column or row is faulty, the array with faults is unrepairable without D-deletion, the signal "1" indicating so is output from the terminal UNRP and the D-deletion process is executed once. Then go to 1). Otherwise, go to the next step.
This step corresponds to Step III in EDR-ALG.

3. The array is repairable and if $n_f = 0$ for all columns and rows, the repairing process is ended. Otherwise, go to the next step.

4. A spare PE by which each unrepaired faulty PE in closed cycles in Step 5 in DR-ALG will be replaced is determined as follows.
This step corresponds to Step IV in EDR-ALG.
To begin with the leftmost column, do the following.

(i) Check whether there are unrepaired faulty PEs in the column. This is done by sending a signal "1" from the lowest row in the column toward the upper. If it is confirmed that there is none, go to the column next to the right. Otherwise, there are unrepaired two faulty PEs in the column and one of them located in the lower row, say PE A, than another PE B will receive the signal "1". Then, PE A sends signal "1"s to the left and right and is replaced by a spare PE located in the left side.
As a general rule,

- A faulty PE which has received a signal "1" from the upper or lower sends signal "1"s to the right and left, and is replaced by a spare PE located in the left side.
- A faulty PE which has received a signal "1" from the left or right sends signal "1"s to the upper and lower, and is replaced by a spare PE located in the upper side.
- A healthy or repaired PE only passes the signal which it has received.

(ii) Finally, a signal "1" must reach the PE A via PE B. If the column checked is the rightmost column, this process is ended. Otherwise, go to the column next to the right and go to (i).   □

First, we show a logical circuit NET-1 which realizes from 1) to 3) in the detailed outline. Next, we show a logical circuit NET-2 which decides the directions of replacements for faulty PEs in closed cycle in 4) in the detailed outline.

Figure 11 shows the logical circuit NET-1 which consists of modules MPE, MSP, two shift-registers each with the same function and several gates, where SP is a spare PE, MPE is shown in (b), and MSP in (c). We explain the functions of the modules in the following.

Assumption:

- Each PE (including spare PE) outputs 1 as it's fault signal if it is faulty, and 0 otherwise. The fault signal of a nonspare PE (spare PE) is input to the terminal F of MPE (MSP) as shown in Fig. 11.

Notation:

- $PE(x, y)$ (including spare PE) denotes the PE in the x-th row and y-th column, where $PE(0, y)$ ($PE(x, 0)$) denotes the spare PE on the upper (left) side of an array.
- Input signal to a terminal number n is denoted as i_n. The output signal out of a terminal number m is denoted as o_m.
- $i(x, y)$ and $o(x, y)$ denote the input and output to/from $MPE(x, y)$, where $MPE(0, y)$ ($MPE(x, 0)$) denotes MSP in the y-th column (x-th row).
- $n_f(0, y)$ ($n_f(x, 0)$) denotes the number of unrepaired faulty PEs including the spare PE in the y-th column (x-th row).

- The function of C_1

  C_1 is used to count n_f in a row or a column and check whether $n_f > 2$. To do so, C_1 adds binary numbers $(x_1 x_0)_2$ and $(0f)_2$, and outputs a binary number $(y_1 y_0)_2$ but the sum is upper-limited to 3, that is, $(11)_2$ if it is greater than 2, as shown in Fig. 11(a).

- The function of MPE

  1. If both the signals i_5 and i_8 are 1s (so initially as will be shown in 1) of the behavior of NET-1), $o_f = i_F$, and 0 otherwise.

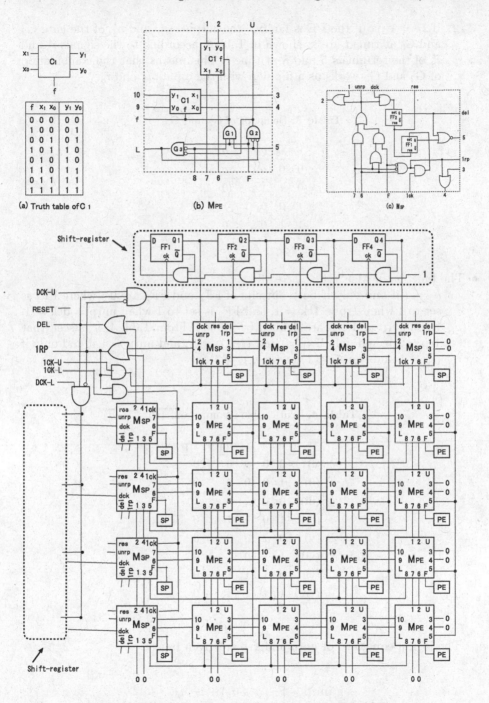

Fig. 11. NET-1 for executing Steps 2 to 5 in DR-ALG.

2. If $i_F = 1$ (so if the PE is faulty), the outputs o_U^t and o_L^t of the gate G_2 and G_3 at time t are as shown in Table 3, according to the signals i_5^t and i_8^t of the terminals 5 and 8 at time t. This means that the combination of G_2 and G_3 works as a flip-flop with the inputs i_5 and i_8.

Table 3. Behavior of G_2 and G_3

i_5^t	i_8^t	o_U^t	o_L^t
0	0	$o_U^{(t-1)}$	$o_L^{(t-1)}$
0	1	1	0
1	0	0	1
1	1	0	0

- The function of MSP

  $o_5 = 1$ if and only if both outputs of FF_1 and FF_2 are 0s, where FF_1 is set to 1 when $1rp = 1ck = 1$, and FF_2 is set to 1 when $unrp = dck = 1$. Here, $unrp$ and $1rp$ are defined by Table 4. From Table 4, it is seen that $1rp = 1$ if and only if $i_F=0$ and $(i_7 i_6)_2 = 1$, and $unrp = 1$ if and only if $(i_F = 1$ and $(i_7 i_6)_2 \geq 1)$ or $((i_7 i_6)_2 \geq 3)$.

Table 4. Truth table for unrp and 1rp

	F	i_7	i_6	unrp	1rp
(i)	0	0	0	0	0
(ii)	1	0	0	0	0
(iii)	0	0	1	0	1
(iv)	1	0	1	1	0
(v)	0	1	0	0	0
(vi)	1	1	0	1	0
(vii)	0	1	1	1	0
(viii)	1	1	1	1	0

The logical equations of unrp and 1rp are given by

$$unrp = F \cdot (i_7 + i_6) + i_7 \cdot i_6.$$
$$1rp = \bar{F} \cdot \bar{i}_7 \cdot i_6.$$

(The behavior of circuit NET-1)

1. Initially, all the flip-flop FFs in MSP's and the shift-registers are reset; that is, o_5 and dck of each MSP is 1 and 0, respectively. Then, i_5 and i_8 of each MPE are 1s, and $(i_7 i_6)_2$ of each MSP shows n_f in the row or column though it is upper-limited to 3.

2. The behavior of NET-1 is controlled by the clocks input to the terminals 1CK-U, 1CK-L, DCK-U and DCK-L. They are fed as follows, which is called D-process.

   D-process:
   (i) While 1RP = 1, 1CK-U and 1CK-L are fed alternatively as $1ck$-$u(1)$, $1ck$-$\ell(1)$, $1ck$-$u(2)$, $1ck$-$\ell(2)$, ... where $1ck$-$u(i)$ and $1ck$-$\ell(i)$ are the i-th clocks to 1CK-U and 1CK-L, respectively.
   (ii) If UNRP = 0 then D-process is ended. Otherwise, DCK-U is fed. Then, if 1RP becomes 1, while 1RP = 1, 1CK-U and 1CK-L are fed alternatively.
   (iii) If UNRP = 0 then D-process is ended. Otherwise, DCK-L is fed. Then, if 1RP becomes 1, while 1RP = 1, 1CK-U and 1CK-L are fed alternatively as in (i).
   (iv) If 1RP becomes 0, go to (ii).

3. If 1RP = 1, there is an MSP whose output o_{1rp} is 1. When 1RP = 1 and a clock through 1CK-U is input to 1ck of MSP ($=$ MPE$(0, y)$), if 1rp from the MSP is 1, i.e., PE$(0, y)$ is healthy and $n_f(0, y) = 1$, FF$_1$ of the MSP is set to 1 and $o_5(0, y)$ becomes 0. Then, o_U of MPE(x, y) with faulty PE(x, y) becomes 1. This indicates that the faulty PE(x, y) is replaced by the spare PE in the y-th column. Further, o_U's and o_L's of all the other MPE's in the y-th column become 0's because i_F of the other MPE's in the y-th column are 0's, $i_5 = 0$ and $i_8 = 1$.

4. When a clock through 1CK-L is input, the similar move to that in above 3) is performed where "column" is replaced by "row".

5. DEL $=. 1$ if and only if unrp of some MSP is 1, and unrp of some MSP is 1 if and only if $n_f \geq 3$ or ($n_f = 2$ including a faulty spare) in a row or a column. Hence, DEL = 1 of NET-1 indicates that the array with the faults is irreparable without D-deletion.

6. If DEL = 0, unrp's of all MSP's are 0's, which indicates that the array with the faults is repairable. Then, if there is neither row nor column such that $n_f = 2$, the spare is healthy and o_5 of MSP is 1, this repairing D-process can be successfully ended with or without D-process. Then go to the process for finding closed cycles (even if there may not be such cycles) together with the directions of replacing for faulty PEs in the cycles. This process corresponds to 4) in the detailed outline and is executed by NET-2 shown in Fig. 13.

- The function of M+
  1. The terminal f of MPE in NET-1 is connected to the terminal F of M+.
  2. If i_F ($= o_f$) is 0, the internal structure becomes as shown in Fig. 13(b), i.e., the signals pass through horizontally and vertically.
  3. If $i_F = 1$, the internal structure becomes as shown in Fig. 13(a). Then,

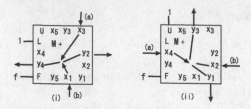

Fig. 12. Signal flow of inputs and outputs when $i_F = 1$

(i) (a) the signal through x_3 from the top or (b) x_1 from the bottom are transferred to the left through y_4 and the right through y_2, and the signal is stored in the flip-flop FFL which indicates that the direction of replacement is to the left. This scene is seen in (i) in Fig. 12.

(ii) (a) The signal through x_4 from the left or (b) x_2 from the right are transferred to the lower through y_1 and the upper through y_3, and the signal is stored in the flip-flop FFU which indicates that the direction of replacement is to the upper. This scene is seen in (ii) in Fig. 12.

- The behavior of circuit NET-2

Note that (i) the internal structure of M+ becomes as shown in (b) in Fig. 13 if a PE with the M+ is healthy or has been repaired and in (a) if it has not yet been repaired, and (ii) there are exactly two unrepaired faulty PEs in a row or column in a closed cycle.

1. Initially, all the flip-flops are reset.
2. Signal "1" is shifted from the left to the right in the shift-register at the time when the output of G_1 is 0 and a clock pulse is fed to CLK-1.
3. Increasing i from 1 to X, the following is performed.
   (i) A clock is fed to all the flip-flops in Fig. 13 through CLK-2 except ones in the shift-register.
   (ii) If Q_i of the shift-register becomes 1, this signal "1" is input to x_1 of M_+ in the bottom row of i-th column. At the time, the output of the gate G_1 becomes 1 and hence a clock to CLK-1 is inhibited to be supplied to the shift register. While a clock to CLK-1 is not supplied, the signal "1" input to the i-th column behaves as follow.
      - If there is no unrepaired faulty PE in the i-th column, the signal "1" passes through all the M_+s in the column, turns back at the M_+ in the top row of the column (note that the terminals x_5 and y_3 are connected) and reaches the terminal y_5 of the M_+ in the bottom row of the column.
      - If there are unrepaired faulty PEs, there are exactly two such PEs in the column whose M_+s are denoted as M_+^L and M_+^U where the former is in a lower location. The signal "1" is fed to M_+^L and propagates in a closed cycle as mentioned in 4) in the detailed outline of hardware realization, finally reaches M_+^U, sent in the upper direction, turns back at M_+ in the top row and reaches the terminal y_5 of the M_+ in the bottom row of the column.

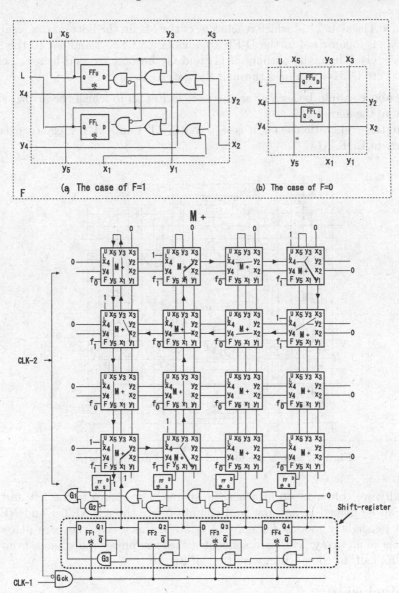

Fig. 13. NET-2 for deciding the directions of replacement while executing Step 6 in DR-ALG

- The signal "1" which reaches y_5 of the M_+ in the bottom row as above is memorized in the D-FF by a clock to CLK-2 and fed to the gate G_2. Then, the outputs of G_2 and G_1 become 0s, and hence, a clock to CLK-1 can pass through the gate G_{ck}.

The above should be seen as an instance in Fig. 13 in which the arrows show the flow of the signal "1".

From the explanation so far, it is seen that NET-1 and NET-2 exactly execute each step in EDR-ALG.

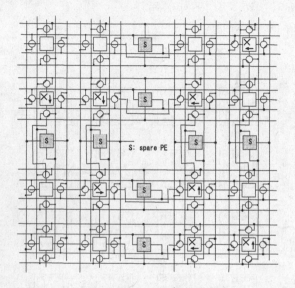

Fig. 14. Spares are located in the center of an array

Finally, we comment on locating the spares. The spares in the left and the upper can be placed in any row and column freely though NET-1 and NET-2 need to be slightly modified. Figure 14 is an instance that they are placed in the center of an array. This arrangement makes the physical distances from the spare PEs half, in comparison with that of Fig. 10.

6   Conclusion

It has been shown that combining the redundancy and the degradation approaches, system lifetime extremely increases. Further, it has been shown that the proposed approach can be realized by the digital circuits consisting of comparatively simple modules which can easily be embedded in a target PA to recover from the fault without the aid of a host computer for reconfiguration. This implies that the proposed method is so useful in enhancing especially the run-time reliabilities and availabilities of PAs in mission critical systems where first self-reconfiguration is required without an external host computer or manual maintenance operations.

References

1. Schaper, L.W.: Design of multichip modules. Proc. IEEE **80**(12), 1955–1964 (1992)
2. Okamoto, K.: Importance of wafer bonding for the future hype-miniaturized CMOS devices. ECS Trans. **16**(8), 15–29 (2008)
3. Dally, W.J., Towles, B.: Route packets, not wires: on-chip interconnection networks. In: Proceedings of the 38th Design Automation Conference, pp. 684–689, March 2001
4. Kung, S.Y., Jean, S.N., Chang, C.W.: Fault-tolerant array processors using single-track switches. IEEE Trans. Comput. **38**(4), 501–514 (1989)
5. Mangir, T.E., Avizienis, A.: Fault-tolerant design for VLSI: effect of interconnection requirements on yield improvement of VLSI designs. IEEE Trans. Comput. **c-31**(7), 609–615 (1982)
6. Negrini, R., Sami, M.G., Stefanelli, R.: Fault-Tolerance Through Reconfiguration of VLSI and WSI Arrays. MIT Press Series in Computer Systems, MIT Press, Cambridge (1989)
7. Koren, I., Singh, A.D.: Fault tolerance in VLSI circuits. IEEE Comput. **23**(7), 73–83 (1990)
8. Roychowdhury, V.P., Bruck, J., Kailath, T.: Efficient algorithms for reconstruction in VLSI/WSI array. IEEE Trans. Comput. **39**(4), 480–489 (1989)
9. Varvarigou, T.A., Roychowdhury, V.P., Kailath, T.: Reconfiguring processor arrays using multiple-track models: the 3-tracks-1-spare-approach. IEEE Trans. Comput. **42**(11), 1281–1293 (1993)
10. Fukushi, M., Fukushima, Y., Horiguchi, S.: A genetic approach for the reconfiguration of degradable processor arrays. In: IEEE 20th International Symposium on Defect and Fault Tolerance in VLSI Systems, pp. 63–71, October 2005
11. Fukushima, Y., Fukushi, M., Horiguchi, S.: An improved reconfiguration method for degradable processor arrays using genetic algorithm. In: IEEE 21st International Symposium on Defect and Fault Tolerance in VLSI Systems, pp. 353–361, October 2006
12. Wu, J., Zhu, L., He, P., Jiang, G.: Reconfigurations for processor arrays with faulty switches and links. In: 15th IEEE/ACM International Symposium on Cluster, Cloud and Grid Computing, pp. 141–148 (2015)
13. Qian, J., Zhou, Z., Gu, T., Zhao, L., Chang, L.: Optimal reconfiguration of high-performance VLSI subarrays with network flow. IEEE Trans. Parallel Distrib. Syst. **27**(12), 3575–3587 (2016)
14. Qian, J., Mo, F., Ding, H., Zhou, Z., Zhao, L., Zai, Z.: An improved algorithm for accelerating reconfiguration of VLSI array. Integration VLSI J. **79**, 124–132 (2021)
15. Ding, H., Qian, J., Huang, B., Zhao, L., Zai, Z.: Flexible scheme for reconfiguring 2D mesh-connected VLSI subarrays under row and column rerouting. J. Parallel Distrib. Comput. **151**, 1–12 (2021)
16. Kuo, S.Y., Chen, I.Y.: Efficient reconfiguration algorithms for degradable VLSI/WSI arrays. IEEE Trans. Comput.-Aided Des. **11**(10), 1289–1300 (1992)
17. Low, C.P., Leong, H.W.: On the reconfiguration of degradable VLSI/WSI arrays. IEEE Trans. Comput. Aided Des. Integrat. Circ. Syst. **16**(10), 1213–1221 (1997)
18. Takanami, I., Horita, T., Akiba, M., Terauchi, M., Kanno, T.: A built-in self-repair circuit for restructuring mesh-connected processor arrays by direct spare replacement. In: Gavrilova, M.L., Tan, C.J.K. (eds.) Transactions on Computational Science XXVII. LNCS, vol. 9570, pp. 97–119. Springer, Heidelberg (2016). https://doi.org/10.1007/978-3-662-50412-3_7

Structural Composite Feature Triangulation
for Visual Object Search

Sinduja Subbhuraam$^{(\boxtimes)}$ (iD)

School of EEE, Nanyang Technological University, Singapore 639798, Singapore
sinduja001@e.ntu.edu.sg

Abstract. This paper proposes a new Structural Composite Feature Triangulation (SCFT) framework for Visual Object Search. Visual Object Search methods often use feature matching approaches which adopt Bag-of-Words (BoW) image representation and employ Geometric Verification (GV) to check the geometric consistency of the putative matching feature pairs between the query and the database image. However, these approaches are unable to handle multiple object instances in an image and they do not utilize the joint local structural information of the matching feature pairs to perform visual search. In view of this, this paper proposes the SCFT method to address these shortcomings. Its key contributions center on the formulation and proposal of a Structural Composite Feature Triangulation method to mine and detect composite structures from Delaunay Triangulation. The composite structures perform detection and localization of multiple object instances within an image. Experimental results on the Belgalogos dataset [1] and the Traffic sign dataset [2] demonstrate that the proposed method can detect multiple object instances, and the top retrieved images show that the proposed method is able to effectively retrieve the relevant images. The results show that SCFT achieves around 4% increase in mean Average Precision (mAP) when compared with that obtained by state-of the-art methods for the Belgalogos dataset.

Keywords: Composite triangulation · Object localization · Visual words

1 Introduction

Over the last decade, Visual Object Search based on the Bag-of-Words (BoW) approach has demonstrated good retrieval performance [1,3–27]. Geometric Verification (GV), an important post-processing step in the Bag-of-Words (BoW) based object retrieval, has shown to improve the retrieval performance of the object search system [1,3–15]. In the BoW approach, the local features obtained from the query and the database images are quantized into visual words. Due to the effects of quantization, the initial set of feature matches between the query and the database features consists of unreliable correspondences. Geometric Verification checks the geometric consistency of the feature matches to eliminate the false correspondences.

This research was partially carried out at the Rapid-Rich Object Search (ROSE) Lab at the Nanyang Technological University, Singapore.

© Springer-Verlag GmbH Germany, part of Springer Nature 2022
M. L. Gavrilova and C. J. K. Tan (Eds.): Trans. on Comput. Sci. XXXIX, LNCS 13460, pp. 22–43, 2022.
https://doi.org/10.1007/978-3-662-66491-9_2

Several GV techniques exist in the literature. Random Sample Consensus (RANSAC) [6] is a typical GV technique used with the BoW model [1,3–15]. It estimates the transformation between the query and the database features belonging to the correspondences to eliminate the false matches. However, RANSAC is time consuming, hence limits the inclusion of large number of images for verification. To address this, a number of techniques [16–27] have been proposed to fasten the GV process. The techniques [16–23] usually exploit the orientation, scale, and location consistency of the feature matches to find the true matches. They either deal with individual correspondences or pairs of correspondences. The other methods [24–27] have adopted Delaunay Triangulation. However, the methods [16–27] do not capture well the similar geometric structures between the query and the database image which are highly beneficial as these structures can well define the similarity between the query and the database image and also help in the detection of multiple instances of an object with varying dimensions in an image.

In this paper, a new Structural Composite Feature Triangulation (SCFT) based Geometric Verification technique has been proposed to detect similar composite structures between the query and the database image. Firstly, the proposed method finds potential subsets of reliable orientation-consistent feature pairs between the query and the database image. Such subsets of true feature pairs help in finding reliable similar structures between the query and the database image. The SCFT utilizes the joint local structural information of the orientation-consistent feature pairs to mine and detect composite geometric structures from the Delaunay Triangulation. Each composite structure, consisting of geometrically consistent adjacent delaunay triangles, helps in identifying the similar local structures between the query and the database image. The detected composite structures are used to localize multiple instances of an object with varying dimensions in an image.

2   Literature Review

As mentioned above, a number of fast GV techniques [16–27] exist. Weak Geometric Consistency (WGC) [16] is one of the techniques which exploits the orientation and scale consistency of the feature matches to find the true matches. Though WGC has shown to speed up the GV process [16–18], it makes use of the geometric (scale and orientation) relations of the individual correspondences; therefore, it does not check for the structural similarity between the query and the database features. Capturing local structures among the features is highly beneficial as these structures can well define the similarity between the query and the database image. The enhanced WGC [19] is another method which includes translation in addition to the orientation and the scale information for consistency check. It integrates orientation, scale, and translation information and generates one histogram based on translation for similarity re-ranking. However, it still works on the individual correspondences. Unlike enhanced WGC which reduces the two dimensional translation vector into a single L-2 norm value, Strong Geometric Consistency (SGC) [20] makes use of the original 2D translation vector. It is based on the assumption that the feature points with similar translation must have similar scale and rotation changes. It divides the matched feature pairs into groups based on their rotation angles and finds the dominant translation for each group. It then uses the number of similarly translated pairs in the largest group to represent the similarity between the two images. The 2D translation vector deals with the pairs of correspondences and adds in more information than that obtained from individual correspondences. Fast Geometric Re-ranking (FGR) [21] adopts the orientation/scale/location based geometric scoring method to re-rank the images. Similar to WGC, the orientation and the scale based scoring deal with the individual correspondences. The location based similarity scoring in FGR analyzes the pairwise distance ratio relations to bring in more discriminative information between the query and the database image. The ensemble of weak geometric relations [22] analyzes neighborhood scaling, rotation, normalized, and relative position coherence of feature pairs to determine true correspondences. Pairwise Geometric Matching (PGM) [23] analyzes the orientation and scale relations of the individual correspondences as well as the pairs of correspondences. Though the pairwise relations provide more information to identify geometrically correct matches, they still cannot capture well the similar geometric structures between the query and the database image. Subsequently, this would result in decrease in the retrieval performance. Due to the non-identification of reliable local structures, the existing fast GV methods cannot deal with the localization of multiple instances of an object with varying dimensions in the database image.

Further, several existing methods have adopted Delaunay Triangulation (DT) for fast Geometric Verification. The method proposed in [24] locally groups features into triplets using multi-scale Delaunay Triangulation, represents the triplets by signatures, and stores them in an inverted file. Each triangle consists of three local features corresponding to three visual words. A signature is generated as an ordered triplet of the three visual word labels of the triangle features, in lexicographically ascending order [24]. Similarly, signatures are generated for all query triangles and compared with the signatures in the inverted index to identify identical signatures which determine the similarity score. Another method [25] utilizes the edges and the triangles formed from the local features to find the true pairs. In addition to recording the three visual words of a triangle as in [24], it also records the angle between the edges and the relative orientations of the triangle. This helps in capturing the shape of the triangle. However, both the methods [24, 25] work on the single triangle geometry which cannot reveal the similar composite structures that exist between the database and the query images. The other methods such as [26, 27] also adopt the Delaunay Triangulation method. The method proposed in [26] works on a simple idea that the putatively matched keypoints that have other putatively matched keypoints in neighborhood are more likely to be true matches. They apply DT on putatively matched keypoints and define the neighborhood of a keypoint as the keypoints that are connected to it by an edge. The number of putative matches among the neighbors of a keypoint is taken as the score. A match with the higher score gets higher priority for inclusion in the list for RANSAC verification. The method proposed in [27] constructs the Delaunay triangles and utilizes the stable properties of Delaunay Triangulation that ensure if two images are similar, the two triangulations should share many triangles with the same vertices [27]. Since both the methods do not adopt an initial geometric consistency check (such as use either orientation/scale/location check), more false pairs exist which affect the DT formation and further processes. Moreover, the methods don't analyze the composite structures of triangles which are even more beneficial to find true pairs.

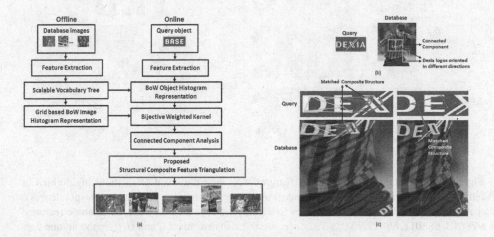

Fig. 1. (a) Visual Object Search framework with the proposed Structural Composite Feature Triangulation. (b) Query image (Dexia logo) and the database connected component (green colored rectangular connected boxes on the database image). (c) Matched composite structures detected between the query image and the database connected component. For e.g., magenta colored query and database structure is a matched composite structure. (Color figure online)

3   Proposed Framework

A typical BoW-based Visual Object Search System [15] with the proposed Structural Composite Feature Triangulation (SCFT) is shown in Fig. 1(a). During the offline phase, the local features are extracted from the database images. The local features contain information regarding their location, affine parameters, and feature description. These features are then quantized into visual words using the Hierarchical k-means (HKM)/Scalable Vocabulary Tree (SVT) [7]. As in [15], the Grid based Bag-of-Words (BoW) image histogram representation for each database image is then computed. During the online phase, the features of the query object are extracted and quantized into visual words. The BoW histogram for the query object is obtained. The Bijective Weighted Kernel (BWK) proposed in [15] is used to measure the similarity between the database and the query BoW histograms. As in [15], the Connected Component Analysis (CCA) is then applied to obtain connected components. An initial shortlist is obtained using the connected component similarity score. To each of the shortlisted database images, the proposed Structural Composite Feature Triangulation is applied to validate the initial set of correspondences. The SCFT mines similar composite structures from subsets of orientation-consistent feature pairs between the query and the database image. Figure 1(b) shows the query image Dexia and the connected component on the database image obtained using [15]. The database connected component shows

Fig. 2. (a) The database and the query features with their dominant orientations highlighted in red (belonging to unique correspondences) and blue (belonging to repeated correspondences). (b) Unique visual word correspondences (red lines) between the query and the database features. $978613, 888011, 593299, 542351, ..., etc.$ are the visual words. U_9 and U_{11} are the unique correspondences existing between the query and the database features mapped to the visual words 593299 and 366339 respectively. For illustration, only few visual word correspondences are shown. (c) Illustration of Repeated visual word correspondences (blue lines) between the query and the database features. The visual words are 542575 and 838763. R_7 and R_8 are the repeated correspondences existing between the query and the database features mapped to the visual word 542575. (d) Illustration of repeated (blue) visual word correspondences for multiple instances of Kia logo in the database image. v_{12}, v_{23}, and v_{24} are the visual words. For example, R_{43} and R_{103} are the repeated correspondences between the query and the database features mapped to the visual word v_{24}. The total number of visual word correspondences are 215. For clarity, only few features and visual word correspondences are shown. (Color figure online)

two Dexia logos oriented in different directions. Figure 1(c) illustrates the similar composite structures detected between the query image (Dexia logo) and the database connected component. A matched query and database composite structure is represented using the same color. For e.g., red colored pair of query and database composite structure is a matched composite structure. The detected composite structures help in the localization of multiple object instances in the image (the details of which are discussed in the following sections). All images are best viewed in color. We have chosen images which can best illustrate each step in the proposed method. The brightness and contrast of the images have been enhanced for better viewing. The experimental settings are the same for all the images. Database connected component is mentioned as database image in the following sections.

4   Unique and Repeated Correspondences

The initial set consists of unique and repeated visual word correspondences between the query and the database image. A unique correspondence occurs when a single query feature and a database feature are mapped to a visual word. Such an one-to-one relationship signifies that the unique correspondences suffer less from the quantization effects of the vocabulary tree. Hence, they are more reliable and contain potential number of true correspondences. The repeated correspondences represent many-to-many relationship between the query and the database features mapped to a visual word. The existence of repeated correspondences is due to the presence of repetitive features (which is quite common for objects having similar textures or structures) that are assigned to the same visual word or due to the effects of quantization. Figure 2(a) illustrates the database and the query features with their dominant orientations. The red and the blue colored features belong to the unique and the repeated visual word correspondences respectively. Figure 2(b) shows the unique visual word correspondences highlighted in red lines. 978613, 888011, 593299, ..., etc. are the visual words. For example, a unique correspondence, U_{11}, exists between the query and the database feature mapped to the visual word 366339. Figure 2(c) shows the repeated visual word correspondences (e.g., the correspondences R_7 and R_8 between the query and database features mapped to the visual word 542575) highlighted in blue lines.

The repeated correspondences found for the Kia Logo in Fig. 2(d) illustrate the one-to-many correspondences (mapping from query to database), a subset of many-to-many, that occur when multiple instances of a query object exist in the database image. For example, true correspondences (e.g., R_{43} and R_{103}) exist between the query and the database features (found for each Kia instance) mapped to the visual word v_{24}. Similarly, true correspondences exist for v_{12} and v_{23}. From Fig. 2, it is observed that the repeated correspondences also have considerable number of true correspondences. Therefore, repeated correspondences must also be taken into consideration; otherwise, there will be under-utilization of the information. Further, as observed from the dominant orientations of the query and the database features in Fig. 2, true correspondences between the query and the database features exhibit consistent orientation difference. Therefore, higher emphasis can be given to database images which have more number of such correspondences. Hence, the proposed method analyzes the orientation consistency (or similarity) of the correspondences and finds potential subsets of orientation-consistent feature pairs from the unique and the repeated correspondences. Each subset exhibits consistent orientation difference. The larger the subset, the higher the possibility of having more number of true correspondences in it. Therefore, reliable similar structures can be found between the query and the database features belonging to these true correspondences.

5　Structural Composite Feature Triangulation

Let $\{U_m\}$ and $\{R_n\}$ be the set of unique and repeated correspondences that exist between the query and the database features. $m = 1, 2, ..., M$ and $n = 1, 2, ..., N$ where M and N is the number of unique and repeated correspondences respectively. Figure 3(a) illustrates the unique (e.g., U_1, U_2) and the repeated visual word correspondences (e.g., R_4, R_{25}). Let $\{\theta_m^U\}$ and $\{\theta_n^R\}$ be the corresponding sets of orientation differences for the unique and the repeated correspondences where $\theta_m^U = \theta_{d,m}^U - \theta_{q,m}^U$ and $\theta_n^R = \theta_{d,n}^R - \theta_{q,n}^R$ respectively. θ_m^U represents the orientation difference between

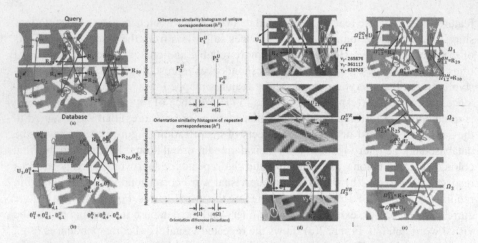

Fig. 3. (a) Unique (U_1, U_2, etc.) and Repeated (R_4, R_5, etc.) correspondences between the query and the database image. For clarity, only few visual word correspondences are illustrated. (b) Computation of orientation differences (e.g., $\theta_1^U = \theta_{d,1}^U - \theta_{q,1}^U$, $\theta_4^R = \theta_{d,4}^R - \theta_{q,4}^R$) for the unique and the repeated correspondences. (c) Orientation Similarity Histograms of the unique and the repeated correspondences. P_1^U, P_2^U, P_3^U, and P_4^U are the local maxima peaks. $\alpha(1)$ is the bin range of P_1^U (includes two peaks on either side of P_1^U). $\alpha(2)$ is the bin range of P_2^U. (d) Illustrates the unique and the repeated correspondences groups (Ω_1^{UR}, Ω_2^{UR}, etc.) belonging to each local maxima peak (along with its two adjacent peaks on either side) on h_U and the corresponding matched peaks on h_R. v_2, v_3 and v_6 are the visual words. (e) Illustrates the Orientation-Consistent Feature Pair subsets (e.g., Ω_1) which contain one-to-one (e.g., $\Omega_{1,1}^{OO} = U_2$, $\Omega_{1,2}^{OO} = R_4$) and one-to-many feature pairs (e.g., $\Omega_{1,1}^{OM} = R_{29}$, $\Omega_{1,2}^{OM} = R_{30}$) selected from the unique and repeated correspondences groups (e.g., Ω_1^{UR}).

the orientations ($\theta_{d,m}^U$ and $\theta_{q,m}^U$) of the database and the query feature belonging to the mth unique correspondence U_m. While, θ_n^R is the orientation difference between the orientations ($\theta_{d,n}^R$ and $\theta_{q,n}^R$) of the database and the query feature belonging to the nth repeated correspondence R_n. Figure 3(b) illustrates the computation of orientation differences for the unique and the repeated correspondences. For example, U_1 is the unique correspondence and $\theta_1^U = \theta_{d,1}^U - \theta_{q,1}^U$ is the corresponding orientation difference between the database orientation ($\theta_{d,1}^U$) and the query orientation ($\theta_{q,1}^U$). Similarly, R_4 is the repeated correspondence between the database and the query feature. θ_4^R is the corresponding orientation difference between the database orientation ($\theta_{d,4}^R$) and the query orientation ($\theta_{q,4}^R$).

The database and the query feature orientations ($\theta_{d,m}^U$, $\theta_{q,m}^U$, $\theta_{d,n}^R$ and $\theta_{q,n}^R$) are obtained as follows. As already mentioned in Sect. 3, each detected local feature contains information about the location and the affine transformation parameters. Let A be the 2×2 affine transformation matrix. The Singular Value Decomposition (SVD) [28] is applied on the A matrix and this results in three matrices, namely, U, D and V matrices. $U = \begin{bmatrix} cos\alpha & -sin\alpha \\ sin\alpha & cos\alpha \end{bmatrix}$ and $V = \begin{bmatrix} cos\beta & -sin\beta \\ sin\beta & cos\beta \end{bmatrix}$ matrices are the rotation matrices and $D = \begin{bmatrix} \lambda_1 & 0 \\ 0 & \lambda_2 \end{bmatrix}$ is the diagonal matrix where λ_1 and λ_2 are the scaling parameters. As in [29], the orientation is then determined from UV^T.

$$UV^T = \begin{bmatrix} uv_{11} & uv_{12} \\ uv_{21} & uv_{22} \end{bmatrix} = \begin{bmatrix} cos(\alpha - \beta) & -sin(\alpha - \beta) \\ sin(\alpha - \beta) & cos(\alpha - \beta) \end{bmatrix}. \ tan(\alpha - \beta) = \frac{uv_{21}}{uv_{11}}.$$ The ori-

entation, $\theta = \alpha - \beta = atan2(\frac{uv_{21}}{uv_{11}})$.

5.1 Orientation-Consistent Feature Pair Subsets

The orientation-consistent feature pair subset Ω_j is defined as,

$$\Omega_j = \{\Omega_j^{OO} \cup \Omega_j^{OM} \mid \theta_j^{OO}, \theta_j^{OM} \in \alpha(j)\}, j = 1, 2, ..., J. \tag{1}$$

The feature pair subset (Ω_j) consists of one-to-one orientation-consistent feature pair subset Ω_j^{OO} and one-to-many (query-to-database) orientation-consistent feature pair subset Ω_j^{OM}. θ_j^{OO} represents the corresponding orientation differences of the feature pairs in Ω_j^{OO} and θ_j^{OM} represents the corresponding orientation differences of the feature pairs in Ω_j^{OM}. $\alpha(j) = [\alpha_{min}(j), \alpha_{max}(j)]$ is the bin range of the jth local maxima peak (the bin range includes two adjacent peaks on either side of the local maxima peak) detected on the orientation similarity histogram of the unique correspondences (h^U). J is the number of local maxima peaks detected.

Let h represent the orientation similarity histogram constructed using the orientation differences in $\{\theta_m^U\}$ and $\{\theta_n^R\}$ respectively. Let h^U represent the orientation similarity histogram constructed using the orientation differences in $\{\theta_m^U\}$ and let h^R represent the orientation similarity histogram constructed using the orientation differences in $\{\theta_n^R\}$. h_k^U and h_k^R count the number of orientation-consistent unique and repeated correspondences in the kth histogram bin. Let $\{P_j^U\}$ be the set of local maxima peaks obtained from the histogram h^U. The peaks or peak values in $\{P_j^U\}$ are arranged in descending order. Let $\{b_j^U\}$ be the corresponding bin number or location set of the local maxima peak set $\{P_j^U\}$. Figure 3(c) shows the orientation similarity histograms $(h^U$ and $h^R)$ for the unique and the repeated correspondences with the detected local maxima peaks $\{P_1^U, P_2^U, P_3^U, P_4^U\}$ highlighted in different colors on h^U and the corresponding matched peaks on h^R. For example, P_1^U denotes the dominant local maxima peak (highlighted in orange) detected on h^U. The other orange colored peaks are the adjacent peaks on either side of P_1^U and $\alpha(1)$ is the bin range which includes P_1^U and two peaks on either side of P_1^U. Soft binning (two peaks on either side of each local maxima peak are considered) is adopted to reduce the quantization effects. The corresponding orange colored peaks on h^R are the matched peaks found using the bin range $(\alpha(1) = [\alpha_{min}(1), \alpha_{max}(1)])$. That is, the peaks in h^R within the bin range $\alpha(1)$ are the orange colored peaks. In terms of bin number, b_1^U is the bin number/location of P_1^U and $b_1^U - 2, b_1^U - 1, b_1^U + 1, b_1^U + 2$ are the bin numbers of the adjacent peaks on either side of P_1^U.

The true correspondences exhibit consistent orientation difference. As observed from Fig. 2 and Fig. 3, unique correspondences have potential number of true correspondences; therefore, the proposed method finds the local maxima peaks from its histogram (h^U). The matching strategy between the two histograms enables the selection of repeated correspondences that have similar orientation difference as that of the unique correspondences. Therefore, the probability of obtaining true correspondences from $\{R_n\}$ is higher with the help of the matching strategy. A narrow bin size (in the experiments it is set to 0.1 radians for the Belgalogos dataset and 0.05 for the Traffic sign dataset) is chosen as it ensures finding reliable unique and repeated correspondences. If the number of unique correspondences is less than a certain threshold (set to 3 in the experiments), the local maxima peaks are obtained from h which is the combined version of h^U and h^R.

Let Ω_j^{UR} be the set of correspondences selected from $\{U_m\}$ and $\{R_n\}$ such that their orientation differences in $\{\theta_m^U\}$ and $\{\theta_n^R\}$ are within the specified bin range $\alpha(j)$. Figure 3(d) shows the correspondences in each subset (Ω_1^{UR}, Ω_2^{UR}, etc.). For instance, the unique and repeated correspondences belonging to subset Ω_1^{UR} are highlighted in orange. Further, it is observed that each subset exhibits consistent orientation difference and the dominant subset contains more number of true correspondences. Within each subset, Ω_j^{UR}, the unique and repeated correspondences are further grouped into one-to-one and many-to-many feature pairs. One-to-one pair indicates that the query feature and the database feature occur only once in the subset or the pair occurs equal number of times. If the pair occurs equal number of times, it is considered only once. For example, Fig. 3(d) shows the one-to-one feature pair belonging to U_2 and the one-to-one feature pairs belonging to R_4 and R_{26}. Many-to-many indicates that the query and the database features occur more than once in the subset. The one-to-many (query to database) and many-to-one (query to database) feature pairs are subsets of many-to-many feature pairs (e.g., the query and the database feature pairs mapped to v_2 (one-to-many) and v_6 (many-to-one) in Fig. 3(d)). From the experiments, it was observed that such feature pairs, especially the one-to-many (mapping from query to database) feature pairs were found to be very useful when dealing with multiple instances. For example, in the case of Kia logo in Fig. 2(b), the query feature, assigned to v_{24}, has true correspondences with the database feature detected in each Kia instance. Retaining such one-to-many feature pairs can help in the detection of multiple object instances in the database image. Therefore, the orientation-consistent feature pair subset Ω_j is formed by selecting the one-to-one and one-to-many feature pairs (from query to database) from Ω_j^{UR}. Within the one-to-many feature pairs, a database point mapped more than once to the same query point is not considered. Considering such points resulted in negligible decrease (e.g., around 0.0002 decrease from 0.3534 (see Table 1) obtained by the proposed SCFT for the Belgalogos dataset) in mAP. Also, the rest of the many-to-many feature pairs within each Ω_j^{UR} are ignored as they were found to be not so useful or they affected the formation of composite structures. Figure 3(e) illustrates the one-to-one ($\Omega_{1,1}^{OO} = U_2$, $\Omega_{1,2}^{OO} = R_4$, $\Omega_{2,2}^{OO} = U_{21}$, etc.) and one-to-many ($\Omega_{1,1}^{OM} = R_{29}$, $\Omega_{1,2}^{OM} = R_{30}$, etc.) orientation-consistent feature pairs in the subsets Ω_1, Ω_2. The corresponding orientations belong to θ_j^{OO} ($\theta_{1,1}^{OO} = \theta_2^U$, $\theta_{1,2}^{OO} = \theta_4^R$, $\theta_{2,2}^{OO} = \theta_{21}^U$, etc.) and θ_j^{OM} ($\theta_{1,1}^{OM} = \theta_{29}^R$, $\theta_{1,2}^{OM} = \theta_{30}^R$, etc.). It can be noted that the one-to-one and one-to-many feature pairs come from both unique and repeated correspondences. An example of one-to-many feature pairs coming from unique correspondences are the indices (3, 4) found in the query and the database image for Quick logo (see Fig. 4(b) illustration).

Algorithm 1: Composite Triangulation(Q_j,D_j)

Input : Q_j ,D_j : Q_j and D_j consists of x and y coordinates of the query and the database features respectively.

Output: G_j: j^{th} Composite Triangulation subset.

1 Check Q_j and D_j has atleast three query and database feature points;
2 **Delaunay Triangulation (DT):**
3 $DT_j^Q \leftarrow \mathrm{DT}(Q_j)$;
4 $DT_j^D \leftarrow \mathrm{DT}(D_j)$;
5 **Matched Triangle Pairs:**
6 $M_j = DT_j^Q \cap DT_j^D$;
7 **Triangle Geometric Consistency Check:**
8 **for** each triangle pair (Q,D) in M_j **do**
9      $\phi^D \leftarrow$ Database Triangle Interior angles (ϕ_1^D ϕ_2^D, ϕ_3^D);
10     $\phi^Q \leftarrow$ Query Triangle Interior angles (ϕ_1^Q ϕ_2^Q, ϕ_3^Q);
11     $\gamma = \sum_{k=1}^3 | \phi_k^D - \phi_k^Q) |$;
12   **if** $\gamma >= th$
13         $M_j \rightarrow$ Remove triangle pair;
14   **end**
15 **end**
16 **Composite Triangulations:**
17 $G_j \leftarrow$ Composite Triangulations from matched triangles in M_j;
18 **return** G_j

Further, as observed from Fig. 3(e), taking into account the other potential local maxima peaks, apart from the dominant one, helps in retaining more number of reliable feature pairs (e.g., feature pairs mapped to v_3) from the unique and the repeated correspondences. The inclusion of other potential local maxima peaks allows the proposed method to handle effectively the following cases: an object with features oriented in different directions, multiple instances of an object with varying dimensions, and quantization effects which occur due to the use of a common bin size selection across all query images.

5.2 Mining of Composite Triangulations

The goal of Composite Triangulation is to mine similar composite structures between the query and the database image using the joint local structural information of the orientation-consistent feature pairs. The method connects the geometrically consistent matched delaunay triangles in the query and the database image based on triangle-adjacency. This helps in identifying strong similar local structures between the query and the database image. These local structures/composite triangulations then aid in the localization of multiple object instances in an image. The flowchart in Fig. 4(a) illustrates the steps involved in obtaining the Composite Triangulations using the orientation-consistent feature pairs in the dominant subset Ω_1. As mentioned in the

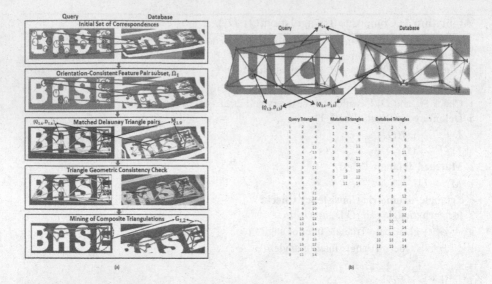

Fig. 4. (a) Flowchart of the proposed Structural Composite Feature Triangulation. For clarity, the first step illustrates only few visual word correspondences with lines. Similarly, the second step also illustrates few feature pairs with lines. From the second step onwards, only the feature pairs in the dominant subset Ω_1 are shown for ease of illustration. From the third step, red colored circles (without the dominant orientation) represent features from the unique correspondences and blue colored circles represent features belonging to the repeated correspondences. $(Q_{1,6}, D_{1,6})$ is an example of a query and database feature pair point. $M_{1,9}$ is a matched triangulation pair. $G_{1,2}$ denotes a matched composite structure. (b) Illustration of the delaunay triangles (green) and the matched triangle pairs (blue, e.g., $M_{1,4}$) in the query (Quick logo) and the database image. $(Q_{1,3}, D_{1,3})$), $(Q_{1,4}, D_{1,4})$ are examples of one-to-many feature pairs from the unique correspondences. Query and Database triangles show the indices of each triangle formed. Matched triangles indicate the indices of the matched triangle pairs. For example, [1 2 4] is a matched triangle pair between the query and the database image. (Color figure online)

previous sections, the first step, the initial set consists of unique and repeated correspondences. The second step identifies the orientation-consistent feature pair subsets Ω_j. For ease of illustration, only the dominant subset Ω_1 is shown. The Matched Delaunay Triangle pairs step (third step) in Fig. 4(a) and Fig. 4(b) illustrate the query, database, and the matched Delaunay Triangulations. Let Q_j and D_j consist of x and y coordinates of the query and database feature pairs in Ω_j. $(Q_{1,6}, D_{1,6})$ in the third step of Fig. 4(a) is an example of a query and database feature pair point. Similarly, $(Q_{1,3}, D_{1,3})$ and $(Q_{1,4}, D_{1,4})$ in Fig. 4(b) are examples of query and database feature pair points. Composite Triangulation algorithm summarizes the sequence of steps to obtain the similar composite structures from the subsets Q_j and D_j respectively. The Composite Triangulation exploits the properties of Delaunay Triangulation (DT) to find the matched triangulation subset (M_j) between the query (DT_j^Q) and the database (DT_j^D) delaunay triangulation subsets. A query and a database triangle is said to be matched if it shares the same delaunay vertices. The query and the database delaunay triangulations are

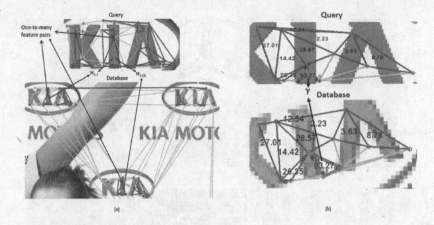

Fig. 5. (a) Illustration of the delaunay triangles (green) and the matched triangle pairs (e.g., $M_{1,1}$ and $M_{1,10}$, highlighted in blue) for multiple instances of Kia logo in the database image. (b) Illustration of geometrically inconsistent matched triangle pairs (pink) with their Geometric Consistency Check measure (γ) values. ϕ_1^D, ϕ_2^D, and ϕ_3^D are the interior angles of the database triangle. ϕ_1^Q, ϕ_2^Q, and ϕ_3^Q are the interior angles of the query triangle. (Color figure online)

highlighted in green and the matched triangulations (e.g., $M_{1,9}$ in Fig. 4(a) and $M_{1,4}$ in Fig. 4(b)) are highlighted in blue. Figure 4(b) shows the indices of the query triangles ([1 2 3], [1 2 4], etc.), the database triangles ([1 2 4], [1 3 6], etc.), and the matched triangles ([1 3 6], [2 5 11], etc.) between them. If there are more than one index (e.g., 3, 4 in the Quick logo) mapped to a point, then the triangles are formed using each index ([1 2 3], [1 2 4], etc.). The matched triangles enable the method to identify similar local structures between the query and the database image. If there are more number of matched triangles, it indicates that the two images share more number of similar structures. The Delaunay Triangulation is sensitive in the presence of false matches. However, the proposed orientation-consistent feature pair subsets (which contains only one-to-one and one-to-many feature pairs) help to reduce a considerable number of false matches. This enables the method to detect more number of reliable matched triangles between the query and the database image. Figure 5(a) illustrates the matched triangle pairs (e.g., $M_{1,1}$ and $M_{1,10}$) for multiple instances of Kia logo in the database image. The use of one-to-many feature pairs enables a single query triangle to find a corresponding match in almost each instance.

Due to the effects of quantization, a feature can be mis-quantized to a different visual word. In such cases, a matched triangle pair may not be geometrically correct (highlighted in pink) as shown in the fourth step, the Triangle Geometric Consistency Check step in Fig. 4(a) and in Fig. 5(b). The Geometric Consistency Check measure (γ) is introduced in the algorithm to identify such triangle pairs. It is based on the assumption that if the sum of the absolute difference between the interior angles of the database (e.g., ϕ_1^D, ϕ_2^D, ϕ_3^D in Fig. 5(b)) and the query (e.g., ϕ_1^Q, ϕ_2^Q, ϕ_3^Q in Fig. 5(b)) triangle is large, then the probability is high for the pair to be geometrically inconsistent. A triangle pair is considered to be geometrically inconsistent if its γ value is greater

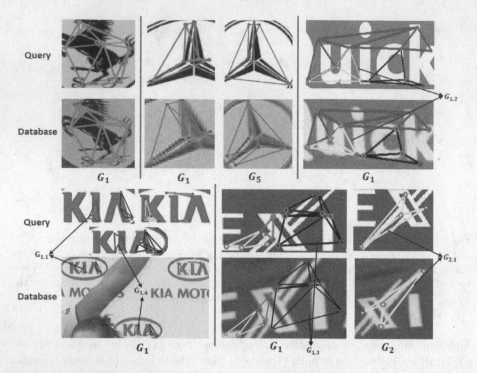

Fig. 6. Composite Triangulations found between the query and the database images. For example, a matched composite triangulation between the query (Ferrari logo) and the database image is highlighted in magenta. G_1 is its composite triangulation subset. For Mercedes logo, G_1 and G_5 are the composite triangulation subsets. $G_{1,2}$ is the composite structure (yellow) in the composite triangulation subset G_1 for Quick logo. For Kia logo, to illustrate the matched structures clearly, the query image is shown separately for each instance found in the database image. $G_{1,1}$ (blue) and $G_{1,4}$ (black) are the composite triangulations belonging to the subset G_1. (Color figure online)

than a certain threshold, th (set to 50 in the experiments). From Fig. 4(a) and Fig. 5(b), it can be observed that geometrically inconsistent pink triangle pairs have large γ values (330.30 and 93.27). The inconsistent triangle pairs are then dropped from M_j.

A group of triangles related to each other provides more information about the local structures of an image than that given by a single triangle. Therefore, the Composite Triangulation is proposed to connect the geometrically consistent triangles found in the database and the query image. The connectivity of the triangles is based on triangle-adjacency. In a Composite structure or triangulation, two geometrically consistent triangles are said to be adjacent if there exists a common edge between them. The matched triangles in M_j are connected resulting in the formation of composite structures G_j in the database and the query image. The fifth step, the Mining of Composite Triangulations in Fig. 4(a) illustrates two matched composite triangulations (highlighted in red and blue) between the query and the database image. $G_{1,2}$ denotes a matched composite

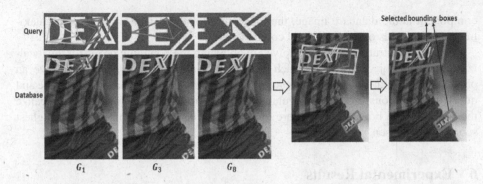

Fig. 7. Illustration of localization of multiple object instances using the composite structures. The intermediate image shows the bounding boxes obtained for each composite structure. The magenta and red colored bounding boxes are the selected boxes indicating the localized regions. (Color figure online)

structure belonging to the subset G_1. It can be observed that the composite structures provide more information about the local structures in the query and the database image.

Figure 6 shows more examples of matched composite triangulations for different pairs of query and database image. Large composite triangulations indicate the existence of strong similar structures between the database and the query image. Such a strong structure can be observed for Ferrari logo wherein almost every triangle shares more than one edge among the other triangles. Further, as can be observed, the use of one-to-many feature pairs (from query to database) helps in identifying more number of similar structures for each Kia instance found in the database image ($G_{1,1}$ and $G_{1,4}$). To illustrate the matched structures clearly, the query image is shown separately for each Kia instance. Finally, the algorithm determines the score (S_j^G) for each subset G_j,

$$S_j^G = w_j \sum_{l=1}^{L} C_{j,l}^2 \tag{2}$$

$C(j, l)$ counts the number of geometrically consistent adjacent matched triangles belonging to the lth composite triangulation found in G_j. L denotes the total number of matched Composite Triangulations found in the subset G_j. The weight $w_j = 2^{-(j-1)}$ is assigned in the order of decreasing importance with the highest weight to the dominant subset G_1 as they contain potential number of true correspondences. The weights are not changed if the local maxima peak values of the subsets are the same. For example, let the set of local maxima peaks from unique correspondences be $\{P_1^U, P_2^U, P_3^U, P_4^U\} = \{12, 2, 2, 1\}$. Then the corresponding weights are $\{2^0, 2^{-1}, 2^{-1}, 2^{-2}\}$ respectively. That is, since P_2^U and P_3^U have same peak values, they have the same weights assigned to them. $S_{l,j} = w_j C^2(j, l)$ denotes the score for each matched composite structure found in G_j. Atleast a single matched triangle is necessary to assign the score for the database connected component. The score for the connected component (S_Ω) in the database image is then obtained as $\sum_{j=1}^{J} S_j^G$. If there are more than one connected component in the database image, the final score for the database image is the maximum of the scores of the connected components.

Figure 7 illustrates the steps involved in obtaining the bounding boxes for the localization of the object instances. Each composite triangulation is used to localize an object. The orientation-consistent feature pairs (x and y coordinates of the database and query triangulations) in each triangulation estimate an affine transform [30] to localize the object. If the bounding boxes are overlapping, the bounding box with the highest score $S_{l,j}$ is chosen.

6   Experimental Results

The proposed method is evaluated on the Belgalogos dataset [1] consisting of 10,000 images and Traffic sign dataset [2] consisting of 48 images. The Bijective Weighted Kernel and Connected Component Analysis (BWK-CCA) visual object search framework [15] is used. The RANSAC part in BWK-CCA is replaced by the proposed Structural Composite Feature Triangulation (SCFT) method. Local features are extracted by the Hessian-Laplace affine detector [31] and represented by the SIFT [32] descriptors. The local features and the descriptors are extracted using the functions in [33] with the default parameters. The descriptors are used to construct a Scalable Vocabulary Tree (SVT) [7] consisting of 1 million visual words for the Belgalogos dataset. For the Traffic sign dataset, the SVT consists of 1000 and 10000 visual words. Also, the Traffic sign dataset is combined with Belgalogos dataset and the SVT is constructed with 100000 visual words (the next paragraph explains in detail the AP and mAP for the two datasets). Similar to [15], the proposed method is applied on each connected component. The algorithms are implemented using VLFeat library [33]. The local maxima peaks are selected such that any two peaks are atleast at some Minimum Peak Distance (in the experiments, it is set to 2) apart. This is done to avoid the selection of less potential smaller peaks whose feature pairs might affect the triangulation formation and localization. The main parameters used in the SCFT are three, namely, Minimum Peak Distance, Bin Size (mentioned in Sect. 5.1), and Geometric Consistency Check measure threshold (mentioned in Sect. 5.2).

The proposed method is evaluated using Average Precision (AP) and mean Average Precision (mAP). Top 100 retrieved images are used for the evaluation of the Belgalogos dataset. Top 20 images are used for the evaluation of the Traffic Road sign dataset. Table 1 shows the AP and the mAP values obtained by RANSAC (th = 5) [15], RANSAC (th = 3) [15], FGR [21], PGM [23], DT [27], and the proposed SCFT for the 11 query logos. The eleven logos are obtained from Google. The visual word vocabulary size is 1 million. For the proposed SCFT, atleast a single matched triangle is necessary to assign a score for the database image or database connected component. Similar to SCFT, the RANSAC part in BWK-CCA is replaced by FGR, PGM, and DT. For the three methods, the threshold is set to 3. That is, atleast 3 database points are needed for the object localization and scoring. Scoring is done as per [21], [23], and [27]. The Orientation geometric scoring method is implemented for FGR. The results in Table 1 show that the proposed method can achieve around 5% improvement over that obtained by RANSAC in [15]. For the method RANSAC (th = 5) in BWK-CCA

Table 1. Retrieval performance comparison of the proposed SCFT with RANSAC (th = 5) [15], RANSAC (th = 3) [15], Fast Geometric Re-ranking (FGR) [21], Pairwise Geometric Matching (PGM) [23], and Delaunay Triangulation (DT) [27] for the 11 query logos in the Belgalogos dataset [1]. RANSAC in [15] uses an inlier threshold (th) of 5. For the proposed SCFT, atleast a single matched triangle is necessary to assign a score for the database image or database connected component. For FGR, PGM, and DT, the threshold is set to 3. Table shows the AP and mAP values for the above mentioned five methods.

LOGOS	RANSAC (th=5) [15]	RANSAC (th=3) [15]	FGR [21]	PGM [23]	DT [27]	Proposed SCFT
BASE	0.469	**0.502**	0.473	0.472	0.450	0.455
DEXIA	0.470	**0.488**	0.476	0.457	0.444	0.471
(logo)	0.040	0.060	**0.092**	**0.092**	0.062	**0.092**
(KIA)	0.522	0.563	**0.566**	0.509	0.552	0.549
(logo)	**0.237**	0.200	0.064	0.128	0.085	0.195
(logo)	0.714	**0.857**	0.425	0.654	0.383	**0.857**
(SNCF)	0.328	0.214	0.372	0.458	0.458	**0.500**
(logo)	0	0.006	0.006	0.011	0.002	**0.014**
(logo)	0.009	0.006	**0.094**	0.068	0.045	0.066
(logo)	0.164	0.113	0.226	0.239	0.182	**0.269**
(logo)	0.330	0.318	0.381	0.336	0.309	**0.420**
mAP	0.298	0.302	0.288	0.311	0.270	**0.353**

Table 2. Retrieval performance of the proposed SCFT for the 3 traffic signs in the I. Traffic sign dataset (Visual Word Vocabulary size = 1000), II. Traffic sign dataset (Visual Word Vocabulary size = 10000), and III. Traffic sign dataset + Belgalogos dataset (Visual Word Vocabulary size = 100000). RANSAC [15] uses an inlier threshold (th) of 5. The Visual Word Vocabulary size (100000) is the same as used in III. Table shows the AP and mAP values for the above mentioned four methods.

LOGOS	I	II	III	RANSAC [15]
(sign)	0.697	**0.824**	0.588	0.588
(sign)	0.524	**0.667**	0.144	0.107
(sign)	0.567	**0.952**	0.410	0.458
mAP	0.596	**0.814**	0.381	0.385

[15], only for the first 6 logos, the AP and mAP values are shown in [15]. For the rest of the logos, the results are shown here. Similarly, for RANSAC (th = 3), the results are shown here. Further, it can be observed that SCFT achieves a 4% increase in mAP when compared with that obtained by FGR, PGM, and DT respectively. Moreover, due to the non-identification of composite structures, FGR, PGM, and DT cannot deal with the localization of multiple instances of an object with varying dimensions in an image. Table 2 shows the AP and the mAP values obtained by the proposed SCFT for the 3 traffic signs in the Traffic sign dataset. The 3 traffic signs are obtained from the dataset itself. For the evaluation, the visual word vocabulary sizes used are 1000 (trained with 48 Traffic sign images), 10000 (trained with 48 Traffic sign images), and 100000 (trained with 10048 images = 48 Traffic sign images + 10000 Belgalogos images). It can be observed that as the visual word vocabulary size increases (from 1000 to 10000), a higher mAP is obtained. The decrease in mAP for the proposed SCFT for the vocabulary size of 100000 is because of the addition of Belgalogos dataset which has considerable

Table 3. Retrieval performance of the proposed SCFT for different values of Minimum Peak Distance. The Belgalogos dataset is used.

Minimum peak distance	0	0.2	0.4	0.6	1	2	2.5	3
mAP		0.352	0.352	0.352	0.352	0.352	0.353	0.353	**0.354**

Table 4. Retrieval performance of the proposed SCFT for different values of bin size. The Belgalogos dataset is used.

Bin size	0.1	0.15	0.2	0.25	3
mAP	**0.353**	0.343	0.341	0.345	0.336

Table 5. mAP values obtained for the local maxima peaks from h and for the local maxima peaks from h^U (proposed in SCFT) for the 11 query logos in the Belgalogos dataset.

	Local maxima peaks from h	Local maxima peaks from h^U
mAP	0.349	**0.353**

number of structures that mimic the three traffic sign queries, namely bike, crossing, and pedestrian. RANSAC [15] uses the same visual word vocabulary size of 100000 as in III. The mAP obtained by RANSAC [15] is almost same as that obtained by the proposed SCFT. Table 3 shows the mAP values obtained by the proposed SCFT for different values of Minimum Peak Distance (MPD) for the Belgalogos dataset. It can be observed that there is not much variation in the mAP for different values of MPD. The slightly lower mAP for MPD of values 0, 0.2, 0.4, etc., is because of the inclusion of smaller peaks which affects the formation of Delaunay Triangulation and the scoring of the triangulations. Table 4 shows the mAP values obtained by the proposed SCFT for different values of bin size for the Belgalogos dataset. From Table 4, it can be observed that a narrow bin size of 0.1 radians has achieved a higher mAP when compared with that obtained by the other bin sizes. A narrow bin size ensures finding groups of reliable unique and repeated correspondences and helps localizing multiple instances of the object with varying dimensions in an image. Table 5 shows the mAP values obtained for the local maxima peaks obtained from the histogram h and for the local maxima peaks from the histogram of unique correspondences h^U (proposed in SCFT). It can be observed that the latter shows a slight increase in mAP when compared with that obtained using the histogram h.

Table 6 and 7 show the mAP values for different histogram peak selection methods for the Belgalogos and the Traffic sign datasets respectively. The first one is the proposed SCFT which uses local maxima peaks from the unique correspondence histogram (2 adjacent peaks on either side of the maxima peak are included as mentioned in Sect. 5.1) to select the potential peaks. This method is compared with the rest of the methods. For all the methods, adjacent peaks are included and the peaks are arranged in descending order.

Table 6. Retrieval performance comparison of the I. Proposed SCFT (local maxima peaks from unique correspondences histogram) with II. Histogram top peak with adjacent peaks and III. Histogram top 3 peaks with adjacent peaks.

	I	II	III
mAP	**0.353**	0.332	0.343

(a)   (b)

Fig. 8. (a) Illustration of the localizations obtained using the proposed SCFT method (local maxima peaks from histogram of unique correspondences). (b) Illustration of the localizations obtained using histogram top peak. For both the peak selection methods, adjacent peaks are included.

The second method uses the top peak from the histogram along with its adjacent peaks. In Table 6, for method II, when compared with method I, it is observed that the mAP is low. This is because the logos have feature pairs whose orientation differences fall in the other potential peaks. The method II does not utilize the other potential peaks. Figure 8(a) shows the localizations achieved by the proposed SCFT and Fig. 8(b) shows the localizations obtained by the second method (histogram top peak). Since the proposed SCFT uses the other potential peaks apart from the top peak, it is able to localize multiple instances of the Dexia logo in the database image. This shows the need to incorporate other potential peaks apart from the top peak. The third method is similar to the second method except that it selects top 3 peaks with its adjacent peaks. A correct localization depends on the true feature pairs present in the selected peaks. Based on the true feature pairs, there is an increase or decrease in the mAP. From Table 6, it is observed that the proposed SCFT achieves a better mAP than the third method. Also, since only the top 3 peaks are selected, as explained for the second method, method III also might not localize some of the object instances.

The same three methods are evaluated on the Traffic sign dataset (Table 7). Method II (which is the histogram top peak selection) achieves an mAP of 0.758 which is the lowest when compared with those of other methods. As mentioned before, this is because of the insufficient number of potential peaks selected. Based on the analysis, it is observed that using the local maxima peaks (proposed in SCFT) achieves a reasonably good mAP for both the datasets.

Figure 9 illustrates the top retrieved images by the proposed SCFT for six different query logos. The visual word vocabulary size used here is 1 million. As observed from Fig. 9, the proposed SCFT is effective in the localization of multiple instances of an object with varying dimensions. This is attributed to the robust detection of similar

Table 7. The evaluation of the three methods mentioned in Table 6 on the Traffic sign dataset. The visual word vocabulary size is 10000.

	I	II	III
mAP	**0.814**	0.758	**0.814**

Fig. 9. Top six retrieved images with the object localization (highlighted in green color) for six different query logos: Kia, Base, Mercedes, Ferrari, President, and Quick. The visual word vocabulary size is 1 million. (Color figure online)

geometric structures between the query and the database image. Figure 10 illustrates the top retrieved images by the proposed SCFT for 3 different traffic signs. The visual word vocabulary size used here is 10000. It can be observed that the proposed SCFT is able to effectively retrieve the relevant images.

The Average computational time obtained by the proposed SCFT is 15.8 s for the eleven query logos in the Belgalogos dataset while for RANSAC [15], it is 27.5 s. The computational time includes query feature extraction, similarity scoring, and object localization. Top 100 images are used for evaluation. The proposed algorithm is implemented using MATLAB on a laptop (2 GHz Intel Core i7-4510U, 16 GB RAM).

Fig. 10. Top six retrieved images with the object localization (highlighted in green color) for three different query traffic signs: bike, crossing, and pedestrian. The visual word vocabulary size is 10000. (Color figure online)

7   Conclusion

In this paper, a new Structural Composite Feature Triangulation has been proposed for robust detection of similar composite structures between the query and the database image. The results show that the proposed SCFT method is able to mine and detect robust Composite Triangulations which help in the localization of multiple instances of an object with varying dimensions. Further, it can be observed that the proposed Structural Composite Feature Triangulation has shown improvement in the retrieval performance when compared with that achieved by RANSAC, FGR, PGM, and DT for the Belgalogos dataset. This shows the benefits of capturing similar structures between the query and the database image for robust Geometric Verification. In the future work, the proposed SCFT will be evaluated on diverse and large scale datasets. Next, Convolutional Neural Network (CNN) features will be integrated with SIFT features to further improve the retrieval performance.

References

1. Joly, A., Buisson, O.: Logo retrieval with a contrario visual query expansion. In: ACM Multimedia, pp. 581–584 (2009)
2. Grigorescu, C., Petkov, N.: Distance sets for shape filters and shape recognition. IEEE Trans. Image Process. **12**(10), 1274–1286 (2003)
3. Xie, L., Tian, Q., Wang, M., Zhang, B.: Spatial pooling of heterogeneous features for image classification. IEEE Trans. Image Process. **23**(5), 1994–2008 (2014)
4. Lampert, C.H., Blaschko, M.B., Hofmann, T.: Efficient subwindow search: a branch and bound framework for object localization. IEEE Trans. Pattern Anal. Mach. Intell. **31**(12), 2129–2142 (2009)
5. Lampert, C.H.: Detecting objects in large image collections and videos by efficient sub image retrieval. In: IEEE International Conference on Computer Vision, pp. 987–994 (2009)
6. Fischler, M.A., Bolles, R.C.: Random sample consensus: a paradigm for model fitting with applications to image analysis and automated cartography. Commun. ACM **24**(6), 381–395 (1981)

7. Nister, D., Stewenius, H.: Scalable recognition with a vocabulary tree. In: IEEE International Conference on Computer Vision and Pattern Recognition, pp. 2161–2168 (2006)
8. Philbin, J., Chum, O., Isard, M., Sivic, J., Zisserman, A.: Object retrieval with large vocabularies and fast spatial matching. In: IEEE International Conference on Computer Vision and Pattern Recognition, pp. 1–8 (2007)
9. Chum, O., Philbin, J., Sivic, J., Isard, M., Zisserman, A.: Total recall: automatic query expansion with a generative feature model for object retrieval. In: IEEE International Conference on Computer Vision, pp. 1–8 (2007)
10. Philbin, J., Chum, O., Isard, M., Sivic, J., Zisserman, A.: Lost in quantization: improving particular object retrieval in large scale image databases. In: IEEE International Conference on Computer Vision and Pattern Recognition, pp. 1–8 (2008)
11. Lin, Z., Brandt, J.: A local bag-of-features model for large-scale object retrieval. In: Daniilidis, K., Maragos, P., Paragios, N. (eds.) ECCV 2010. LNCS, vol. 6316, pp. 294–308. Springer, Heidelberg (2010). https://doi.org/10.1007/978-3-642-15567-3_22
12. Li, T., Mei, T., Kweon, I.S., Hua, X.S.: Contextual bag-of-words for visual categorization. IEEE Trans. Circ. Syst. Video Technol. 21(4), 381–392 (2011)
13. Arandjelovic, R., Zisserman, A.: Three things everyone should know to improve object retrieval. In: IEEE International Conference on Computer Vision and Pattern Recognition, pp. 2911–2918 (2012)
14. Jiang, Y., Meng, J., Yuan, J., Luo, J.: Randomized spatial context for object search. IEEE Trans. Image Process. 24(6), 1748–1762 (2015)
15. Sinduja, S., Yap, K.H., Dajiang, Z.: Bijective weighted kernel with connected component analysis for visual object search. IEEE Sig. Process. Lett. 22(10), 1604–1608 (2015)
16. Jegou, H., Douze, M., Schmid, C.: Hamming embedding and weak geometric consistency for large scale image search. In: Forsyth, D., Torr, P., Zisserman, A. (eds.) ECCV 2008. LNCS, vol. 5302, pp. 304–317. Springer, Heidelberg (2008). https://doi.org/10.1007/978-3-540-88682-2_24
17. Jegou, H., Douze, M., Schmid, C.: On the burstiness of visual elements. In: IEEE International Conference on Computer Vision and Pattern Recognition, pp. 1169–1176 (2009)
18. Jegou, H., Douze, M., Schmid, C.: Improving bag-of-features for large scale image search. Int. J. Comput. Vis. 87, 316–336 (2010). https://doi.org/10.1007/s11263-009-0285-2
19. Zhao, W.L., Wu, X., Ngo, C.W.: On the annotation of web videos by efficient near-duplicate search. IEEE Trans. Multimed. 12(5), 448–461 (2010)
20. Wang, J., Tang, J., Jiang, Y.-G.: Strong geometry consistency for large scale partial-duplicate image search. In: ACM International Conference on Multimedia, pp. 633–636 (2013)
21. Tsai, S.S., et al.: Fast geometric re-ranking for image based retrieval. In: Proceedings of IEEE International Conference on Image Processing, pp. 1029–1032 (2010)
22. Wu, X., Kashino, K.: Robust spatial matching as ensemble of weak geometric relations. In: British Machine Vision Conference, pp. 1–12 (2015)
23. Li, X., Larson, M., Hanjalic, A.: Pairwise geometric matching for large-scale object retrieval. In: IEEE International Conference on Computer Vision and Pattern Recognition, pp. 5153–5161 (2015)
24. Kalantidis, Y., Pueyo, L.G., Trevisiol, M., van Zwol, R.: Scalable triangulation-based logo recognition. In: ACM International Conference on Multimedia Retrieval, pp. 1–7 (2011)
25. Romberg, S., Pueyo, L.G., Lienhar, R., van Zwol, R.: Scalable logo recognition in real-world images. In: ACM International Conference on Multimedia Retrieval, pp. 1–8 (2011)
26. Bhattacharya, P., Gavrilova, M.: DT-RANSAC: a delaunay triangulation based scheme for improved RANSAC feature matching. In: Gavrilova, M.L., Tan, C.J.K., Kalantari, B. (eds.) Transactions on Computational Science XX. LNCS, vol. 8110, pp. 5–21. Springer, Heidelberg (2013). https://doi.org/10.1007/978-3-642-41905-8_2

27. Kong, L.-B., Kong, L.-H., Yang, T., Lu, W.: A fast and effective image geometric verification method for efficient CBIR. In: Sharaf, M.A., Cheema, M.A., Qi, J. (eds.) ADC 2015. LNCS, vol. 9093, pp. 232–243. Springer, Cham (2015). https://doi.org/10.1007/978-3-319-19548-3_19

28. Singular Value Decomposition of a 2×2 Matrix. https://sites.ualberta.ca/~mlipsett/ENGM541/Readings/svd_ellis.pdf

29. Hartley, R., Zissermann, A.: Multiple View Geometry in Computer Vision, pp. 39–40. Cambridge University Press, Cambridge (2004)

30. Zuliani, M.: RANSAC toolbox for Matlab (2008). https://github.com/RANSAC/RANSAC-Toolbox

31. Mikolajczyk, K., Schmid, C.: Scale and affine invariant interest point detectors. Int. J. Comput. Vis. **60**, 63–86 (2004). https://doi.org/10.1023/B:VISI.0000027790.02288.f2

32. Lowe, D.G.: Distinctive image features from scale-invariant keypoints. Int. J. Comput. Vis. **60**, 91–110 (2004). https://doi.org/10.1023/B:VISI.0000029664.99615.94

33. Vedaldi, A., Fulkerson, B.: VLFeat: an open and portable library of computer vision algorithms (2008). https://www.vlfeat.org/

Study of Malaysian Cloud Industry and Conjoint Analysis of Healthcare and Education Cloud Service Utiliztion

Chandrani Singh[1,2], Midhun Chakkaravarthy[3], and Rik Das[4(✉)]

[1] Lincoln University College, No. 12-18, Wisma Lincoln, Malaysia
singh.chandrani@gmail.com
[2] Sinhgad Institute of Management (SIOM), Pune, India
[3] Faculty of Computer Science and Multimedia, Lincoln University College,
Wisma Lincoln, Malaysia
divya@lincoln.edu.my
[4] Programme of Information Technology, Xavier Institute of Social Service,
Jharkhand, Ranchi, India
rikdas78@gmail.com

Abstract. The paper initially addresses Malaysia's readiness concerning the adoption of cloud computing services and provides insights into the country's internet infrastructure and internet users, fixed and mobile broadband deployment, and usage of international bandwidth for improvising cloud services. In addition, the paper investigates the scenario around policy areas, including privacy and security laws, being the chief contributors for the adoption of cloud services across regions and borders. The objective behind conducting the research is to validate the current scenario in Malaysia for the cloud service adoption by healthcare and education segment and identifying the consumer service attribute preferences. By performing the conjoint analysis the research would help in identifying the determinants and prioritizing them based on user preferences with prior use of Multinomial Logit choice model (MNL) and Max Likelihood function. A statistical discussion is also presented concerning the current global market scenario and scenario in the Asia Pacific across the verticals.

Keywords: ICT readiness · Service utilization and governance

1 Introduction

Cloud service adoption in any nation is impacted by the internet infrastructure and bandwidth across the urban-rural segment, The base population and the households, the percentage of literacy, the economic and the industry infrastructure. In addition for the adoption to be continuous and consistent, there are legal, privacy and security guidelines that are to be adhered to [1]. To set the base for cloud adoption and to depict the current scenario concerning service provisioning, it is imperative to provide parameter wise country statistics and relative standing of Malaysia with respect to cloud

M. L. Gavrilova and C. J. K. Tan (Eds.): Trans. on Comput. Sci. XXXIX, LNCS 13460, pp. 44–70, 2022.
https://doi.org/10.1007/978-3-662-66491-9_3

adoption. A comparative evaluation of a few major players in Asia, in context of cloud computing market has also been studied to understand the limitations. With reference to the investigation performed, the data controllers in Malaysia, are required to register with the Government Department, those that constitute communications, banking and finance, insurance, healthcare, tourism and hospitality, transportation, education, direct sales, services, real estate, and utility sectors as designated by the Minister of the Malaysian Communications and Multimedia Commission (MCMC) [2].

In addition Malaysia as a nation ensures compliance and promotion of Malaysian standards, through Ministry of Science Technology and Innovation (MOSTI) and is a participant of ICT standards committee, the Joint Technology Council (JTC-1).Malaysia previously had also a strategy in place for the promotion of cloud services known as the Multimedia Super Corridor (MSC), Malaysia Cloud Initiative (MMCI), enabling government to implement technology neutrality to a considerable extent.

2 Related Works

With respect to Cloud Computing service provision in Malaysia, intellectual property laws for effective enforcement of misappropriation and contravention are limited, hence

Table 1. ICT governance policy status

Laws for ICT governance	Availability
Presence of modern electronic commerce, signature laws and data protection regulation act, in alignment with globally recognized frameworks	Yes
Data breach notification provisions	No
Access to encrypted data	Yes
Malaysian Copyright law	Yes (enforcement is sporadic and attitude stringent to intermediate liability)
Cross border transfer of data/ Law for data localization	Permissible/Not applicable
Cyber security strategies and policies	Not clearly devised
ICT security policy	Yes (Few initiatives under MOSTI)
Cyber crime	Yes(MLAT and ASEAN MLACM)
Protection of trade secrets	Yes but not appropriately codified
Intellectual property law	Yes but limited
Censorship laws	Yes
Usage laws –(products and services) and restrictions	Yes mixed mode favoring both locally owned businesses and seeking international tenders
Data Localization laws	Not applicable
Foreign ICT investments	Substantial progress

not much investment in cloud research and development can be envisaged at this point [3]. Concerning the laws and policies in the context of ICT readiness and governance in Malaysia, Table 1 in the related works depicts the present scenario.

Existing problems in Malaysian Healthcare Systems pertains to majorly the traditional approach adopted that are either paper-based and manual or are purely standalone applications [4].The absence of automation, successively collaboration and integration of applications,has been the prime cause for compromise in the quality of services. Healthcare sector in Malaysia needs the desired automation to provide cost effective and timely services, for which the Information Technology (IT) experts are of the opinion that cloud computing solutions can support the healthcare systems and services effectively [5].

Currently most of the hospitals in Malaysia are using their own standalone Hospital Information Management System (HIMS), and these systems face several problems as listed below [5, 6]:

1. Storage problem: storage and transmission of data with limited cost and high storage cost.
2. Heterogeneity and Compatibility: Transfer of files, conversion across multiple formats and integration across multiple systems (public, private, others etc.) is extremely cumbersome as different hospitals in Malaysia use different systems.
3. Setup cost: the setup cost of the HIMS inclusive of the hardware, software and network infrastructure is extremely expensive.
4. Maintenance issue: Pertaining to issues that are complex and require the intervention of third-party resources making it a costly affair altogether.

Several benefits of cloud services adoption has been perceived by the Malaysian Healthcare Sector. In one of research studies conducted, [7] illustrated that utilization and adoption of cloud services ensures reduction in the operational cost, increase in the performance of on cloud medical units, enables consistency in implementation of regulatory compliance's, and boosts healthcare business development in the global market [5].

In a separate study conducted by [8], results show that the cloud computing provides a connected environment to improve operational efficiency. In addition, use of cloud computing in the healthcare sector ensures access to and availability of unlimited virtual resources, increase in the storage capacity and computing power [8].

Cloud computing can also improve the patient information sharing, by using the virtual servers on a pay per usage model.The study by [9] focuses on the relationship among the health insurance and health care providers, the IT resources that help in determining the adoption of cloud computing service to improve and develop the collaboration and efficiency. A study conducted by [10] provides strategy for implementing the prototype based on cloud computing architecture for HIMS.

A research study conducted by [5] showcases the use of the structural equation modeling approach,that proves that usage of cloud computing services in Malaysian healthcare sector have a strong impact on the efficiency of service provisioning, ensuring greater degree of collaboration among the stakeholders and service providers.

But the challenges that have surfaced with the adoption are as follows: restrained government budget for healthcare spends, lack of the integration between systems and data sharing; security issues faced in migrating to cloud computing [11] and availability of a reliable and robust internet connection [10, 12].

With respect to adoption in the education sector, according to McREA (2009), cloud computing contributes more towards improving teaching and research as compared to earlier IT and software systems. It is said that Malaysian government has taken the first step by implementing cloud computing for federal ministries that is managed by MAMPU (Malaysian Administrative Modernization and Management Planning Unit) for several higher education institutions and universities. According to [13], a migration to cloud by education stakeholders can be formed by developing the following:

- Developing the knowledge base about cloud computing
- Evaluating the present stage of the university
- Experimenting the cloud computing solutions
- Choosing the optimum solution
- Implementation and management of the cloud computing solution

This can help in collaborative editing and management of data on virtual servers and sharing of public, private and hybrid cloud infrastructures and repositories to improvise on productive communication between researchers, students and faculty members [14].

The above service provisioning will enable Ministry of Higher Education (MOHE, Malaysia) to easily monitor and access the resources for higher institutions [15]. Around 40 modules are hosted on cloud right from admission to alumni management and from recruitment to retirement for employees. Also with IT infrastructure acquisition and release and payroll to regulatory compliance on cloud, many public and private universities in Malaysia would be benefited from the adoption.

Conjoint analysis has been used as an optimal market research approach for measuring the value that consumers place on features of a product or service and combines real-life scenarios and statistical techniques [16]. By performing the conjoint analysis, the research would help in identifying relative importance [17] of all consumer attribute preferences and their levels thus assigning due priority during service selection.

3   Data Acquisition Methodology

With respect to the data acquisition and the methodology followed, the research data collected is majorly from the secondary sources i.e. from the websites of government and non-government agencies both public and private.

Study has been conducted by various market research companies to conclude on global and regional (Asia Pacific and Malaysia) cloud computing service provisions from the provider and the supplier perspective. This study also presents a global and regional (Asia Pacific and Malaysia) view by sectors and cloud service level offerings in terms of market realizations. It then narrows down to two primary business verticals under consideration i.e. healthcare and education.

The methodology adopted to estimate the global cloud computing market size pertains to a top down approach [18] as shown in Fig. 1. It primarily constitutes of global market size, geographic segmentation and market size by category and percentage and this approach assisted in providing a global overview pertaining to the adoption of cloud services across Education and Healthcare verticals. The cloud readiness index of Malaysia was also assessed in comparison to the other Asia Pacific (APAC) countries. Additionally, the detailed statistics of internet infrastructure, ICT penetration, international bandwidth and network readiness along with legal, security, privacy and trade policy data were collated from various websites (government, private and market research companies) to conclude on Malaysia's current status with respect to IT, Network and cloud services. Narrowing down to the Education vertical, the data from the websites of the universities and institutions (secondary) was collected to investigate the extent, form and type of cloud service adoption and the market size using a bottom-up approach [18]. Similar approach was followed to extract data from government, semi-government, private healthcare providers and consumers. This helped in the ranking of the utility of services provided by the suppliers with respect to their product offering. This methodology of data collection and validation to estimate the market size and other relevant statistics has been adopted by established market research organizations across the world.

Market size estimation using secondary data source and appropriate segmentation has been carried out with respect to region, sub segment and category for detailed investigation.

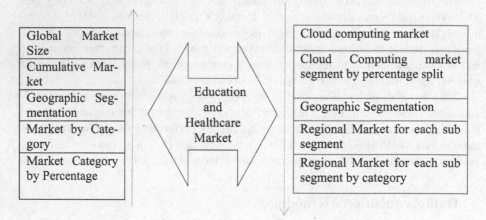

Fig. 1. Top down and bottom up approach for data collection and validation

3.1   Research Data – Secondary Sources

The data segments for the Education and Healthcare verticals tapped and the respective data sources identified are listed in Table 2 below.

Table 2. Data segment and source

Data segment	Data source
Size of the market (Global and Regional)	Published interviews of key opinion leaders of the healthcare and education segment of Malaysia Technology journals Certified publications Market research websites White papers and articles from recognized authors directories, and databases and press notifications
ICT Readiness and Cloud Readiness Parameters (Global)	ITU reports and parameters provided by Asia Cloud Computing Association
Education provider data (K12 and Higher Education) using Cloud solutions (Global and Regional)	Websites and secondary sources for 60 universities, colleges and institutes
Healthcare Provider and Consumer Data (Hospitals, Pharmacies, Imaging centers, Ambulatory etc.) (Global and Regional)	A survey of 135 public hospitals and 9 special medical institutions in Malaysia from secondary sources showcase that the most popular cloud healthcare application offerings is the eHRMIS (Human Resource Management Information System), MyCPD (Continuing Professional Development), and MyHealth websites

3.2 Data Normalization and Aggregation

Having extracted the data to conduct the said research, it was subjected to normalization. Normalization is said to make the data comparable across indicators.

The parameters and the resultant data that have been scaled using different units, and have been brought to a 10 point scale to make indicator values comparable. The normalization was done to compute the cloud readiness index of Malaysia across indicators as cloud governance, security, infrastructure and regulation.

Complex Normalization (Using minimum-maximum method)-This included conversion of absolute values to normal logarithmic values and setting up maximum values to transform the values into scores. Complex normalization was performed to deduce the IT and network readiness of Malaysia. The formula used to calculate the normalized score is denoted in Eq. (1):

$$\text{Normalized Value} = (\text{Acutal} - \text{Min})/(\text{Max} - \text{Min})*10 \qquad (1)$$

3.3 Solution of Missing Data

Treatment of missing values were considered appropriately to avoid skewing of the results by considering countries belonging to the same segment. For example Malaysia belonging to the upper middle economy segment, missing data for some of the parameters was taken from countries in that segment and region and average score was considered for analysis and subsequent interpretation.

4 Research Methodology

The research study establishes the base and the pre-requisites for cloud service adoption in healthcare and education segment in Malaysia. In Phase-I, Global and APAC cloud computing market size and the stage of service provision is delineated to form the base for the study. The research study then narrows down to the IT and Network readiness, Cloud Computing readiness of Malaysia with factors impacting cloud service adoption, along the urban, rural segment and among the stakeholders.

In Phase II the study of Global and APAC Healthcare and Education cloud computing market segment is undertaken. This is then followed by a sector-wise implementation and insight to Malaysia's global and regional standing.

Finally in Phase-III, the study focuses on the market potential and growth across the Healthcare and Education with respect to cloud service adoption in Malaysia.It then performs a conjoint analysis to identify user's attribute preferences in availing cloud services.The cloud services rendered by the popular service providers are considered for the study and the relative importance of attributes considered by the consumers for service selection are presented based on which suppliers are ranked. Multinomial Logit choice model (MNL) and Max Likelihood function was used to deduce the relative importance of attributes deduced with high value of likelihood ratio [13].

4.1 Research Hypothesis

The null hypothesis formulated is that there exists no strictly preferred attributes or consumer preferred attributes and all part worth utilities equal to zero were rejected with significance of $\alpha = 0.01$ and a high likelihood ratio. The researchers from the study conducted found that potential consumers of cloud services do have a strong preference for different service attributes.

Findings and Discussions

The findings and subsequent discussions are enlisted across the phases for both healthcare and education segments with respect to cloud service adoption.

4.1.1 Phase-1

In context to the previous researches conducted, the global cloud computing market size/revenue is expected to grow at a Compound Annual Growth Rate (CAGR) of 17.5% from 2020 to 2025 and APAC market size/revenue up to 117% from 2020 to 2025 [19] as shown in Fig. 2 and 3.

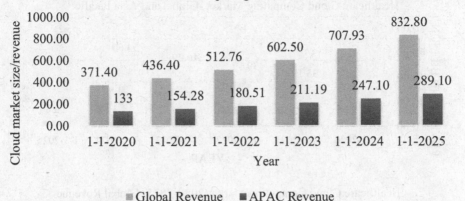

Fig. 2. Global and APAC cloud computing market revenue

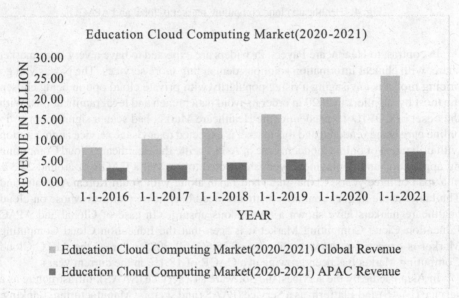

Fig. 3. Education cloud computing market (Global and APAC)

The Global Healthcare Cloud Computing Market is expected to grow to USD 64.7 billion by 2025, at a CAGR of 18.1% year on year as shown in Fig. 4.The growth is attributed to technological upgradation and improvised digitization and deployment of cloud-based Healthcare Information Technology (HCIT) solutions to improve the care process.

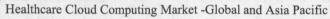

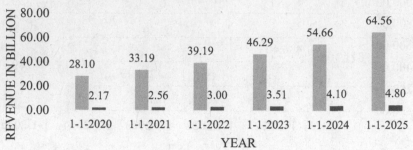

Fig. 4. Healthcare cloud computing market (Global and APAC)

In contrast to Healthcare Payers, Providers are expected to have a very large market share, with clinical information solutions demanding more services. The pay as you go pricing model is envisaging a rising popularity with private cloud option being chosen the most by suppliers in 2020 in order to avoid data breach and legal ramifications. With the onset of COVID 19 pandemic, the Healthcare Market had seen a significant rise in online care being rendered and this market is expected to envisage service augmentation with utilization of online mode. In case of Asia-Pacific the Healthcare Cloud Computing by application and deployment type is expected to grow at a CAGR of around 24.3% in a span of three years. China, India and Japan along with South Korea, Australia and Thailand are the major economies with advanced healthcare facilities whose on-cloud healthcare markets have shown a tremendous upsurge. In case of Global and APAC Education Cloud Computing Market it is seen that the Education Cloud Computing Market is expected to grow at the rate of 25.6% till 2025 whereas the APAC Cloud Computing Market has been growing at a CAGR of 18.1% in the current years.

In Asia Pacific by the services, the Software as a service (SAAS), Infrastructure as a service (IAAS) and Platform as a Service (PAAS) and sectors: Manufacturing, Banking, Financial Services and Insurance (BFSI) and Telecommunications are utilizing the cloud computing adoption the most as researched by [20] and shown in the Fig. 5. Below.

The base indicators for cloud readiness as reported by ICT indicator database presents that Malaysia has envisaged a population increase of 2% from year 2015 with urban segment rising by 0.9% and households by 2%.The ICT indicator database presents the fact that in 2015, 67.6% of households in Malaysia had personal computers which sets an optimism for the adoption of cloud services by various segments and stakeholders across the nation [17].

Continuing with IT and Network Readiness the graph represents the current status of Malaysia with regard to the following:

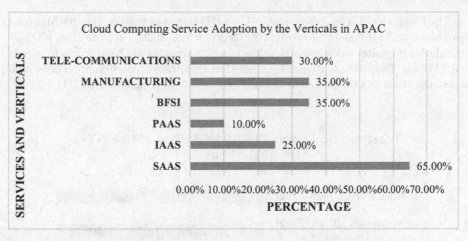

Fig. 5. Cloud computing service adoption by verticals across Asia Pacific

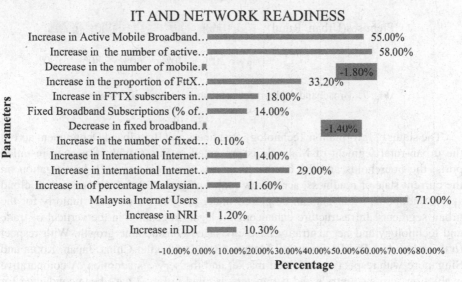

Fig. 6. IT and Network readiness status of Malaysia

- active mobile broadband subscriptions
- increase in the proportion of fibre to home subscribers
- fixed broadband subscription (i.e. percentage of internet users)
- increase in the international internet bandwidth and increase in Malaysian internet users.

The International Telecommunication Union (ITU) report also presents that Malaysia has a National Broadband Plan wherein the target for year 2020 is to ensure that 100% of households in capital cities and high impact growth areas should have access to speeds of 100 Mbps. The 2016–2020 plan for strengthening infrastructure to support economic expansion states that broadband to be deployed in 95% of populated area [21].

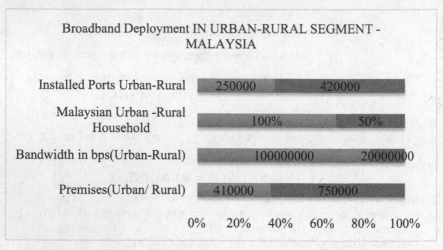

Fig. 7. Broadband deployment in urban-rural segment – Malaysia

The status of Information Technology Readiness and Broadband Deployment across the urban-rural segment of Malaysia is presented in Fig. 6 with respect to installed ports, the households and the bandwidth across the segments [22]. Investigation on the current state of readiness, across few major players of Asia, with respect to cloud service adoption, Malaysia can be placed under progressive category majorly for the urban segments. Infrastructure enhancement, legal procedures in the vertical of trade and technology and net neutrality have envisaged considerable growth. With respect to the other Asia Pacific countries, Malaysia stands 5th after China, Japan, Korea and Singapore with respect to the cloud market and the services adoption. A comparative evaluation across country's and parameters as cited in Fig. 7 have been conducted for improvised clarity.

In succession to the IT and Network Readiness Index, also robust is the cloud infrastructure, security protocols and cloud governance as seen in Fig. 8. The Cloud Readiness Index as reported by Asia Cloud Computing Association, has placed Malaysia at the 8th position in the APAC region as reported in Table 3. The Cloud Regulation score of Malaysia is good owing to the stringent rules and regulations set by the Malaysian Government in accordance with ISO 27018 thus ensuring Privacy and Intellectual Property Protection [22].

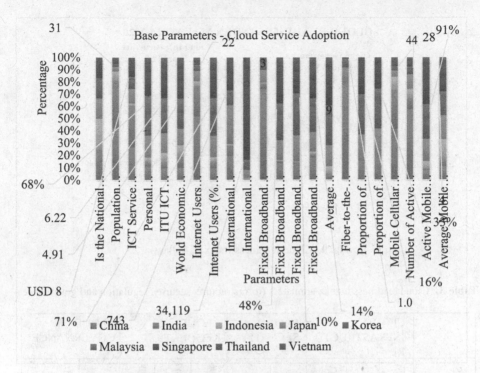

Fig. 8. Comparative study across APAC region

Many APAC economies have announced 'Cloud First' policies and have directed their efforts on making infrastructure, platform and software service implementation a reality. It is recommended by Asia Cloud Computing Association to the emerging economies to focus on supporting initiatives as cloud vendor registration and accreditation, designing of security standards and guidelines, relevant data management policies. It has also stressed on leveraging the potential of technological advancements for economic recovery of emerging Asian markets among which Malaysian markets have been also considered. For Malaysia there has been no change in the rank with respect to Cloud Readiness Index score and it stands till at 8 in the year 2020 as like 2018 with respect to infrastructure, governance, security and regulation (Fig. 9).

With reference to Malaysia as per the ITU report, it is said that public cloud will bring business, with the market expected to grow by 3.5% over 2019.As per the recent reports, also the sector wise impact during COVID 19 across Malaysia is displayed in Fig. 10 and Fig. 11 respectively. Healthcare sector as delineated remains unaffected and ICT spending remaining totally flat. Deferment in the capital expenditures, containment of costs by the enterprises to bring the hardware costs down on one hand and on the other a huge demand for increased internet bandwidth, adoption of cloud-based services and increased usage of collaboration platforms are few of the promising traits exuded.

The next phase i.e. Phase-II of the study encompasses a parameter based scenario across global and local (Malaysia) regions across the healthcare and education segment which forms the fundamental constituent of research study.

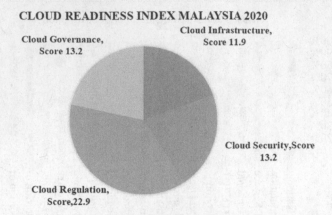

Fig. 9. Cloud readiness Index of Malaysia

Table 3. Cloud readiness index parameters (infrastructure, security, regulation and governance)

CLOUD READINESS INDEX (FACTORS)	CLOUD INFRASTRUCTURE			CLOUD SECURITY		CLOUD REGULATION			CLOUD GOVERNANCE	
	CRI #01	CRI #02	CRI #03	CRI #04	CRI #05	CRI #06	CRI #07	CRI #08	CRI #09	CRI #10
	Int'l Connectivity	Broadband Quality	Power and Sustainability	Data Centre Risk	Cyber security	Privacy	Govt Regulation Environment	IP Protection	Biz Sophistication	Freedom of Information
Score	2.5	5.5	4	4.1	8.9	7.5	7.9	7.6	7.8	5.3
Ranking	10	8	9	8	2	8	3	6	7	11

4.1.2 Phase-II

In this phase a detailed study has been undertaken taking into consideration healthcare and education segment stakeholders perspectives, preferences, cloud implementation readiness at global and regional level.

4.1.2.1 Cloud Computing Market in Healthcare Segment - A Global Preview

Healthcare cloud computing aims at controlling data scalability, storage of clinical statistics in repositories and quicker accessibility to digital medical records and reduced capital

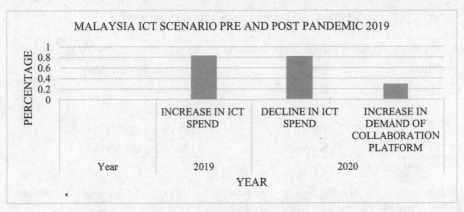

Fig. 10. ICT trends across Malaysia post pandemic 2019

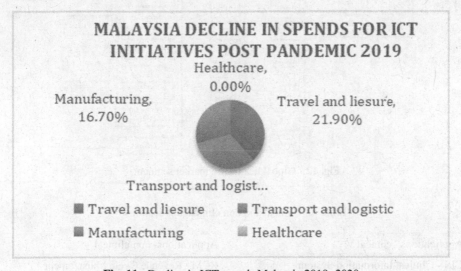

Fig. 11. Decline in ICT spends Malaysia 2019- 2020

expenditure in patients billing. The Healthcare Cloud Computing Market is segmented across Pricing, Services, Deployment, Application, Users, Geography and Component as shown in Table 4 and the market dominance forecast till 2024 across categories is also presented in the succeeding section.

Following are the abbreviations for clinical and non clinical information systems as presented in Table 5 below.

Table 4. Global Healthcare Market Domination

Global Cloud Computing (Healthcare) - Market Domination Forecast (2019–2024)		
2019–2024	Users	Healthcare Providers
	Deployment	Private Cloud
	Component	Services
	Pricing	Pay as you go
	Service	Software as a Service

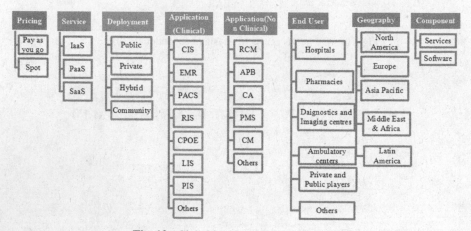

Fig. 12. Global healthcare market segment

Table 5. Clinical and non clinical applications

Applications - clinical	Applications- non clinical
CIS - Clinical Information System	RCM - Revenue Cycle Management
EMR - Electronic Medical Record	APB - Account Payable and Billing
PACS-Picture Archiving and Communication Systems	CA - Compliance and Audit
RIS -Radiology Information System	PMS - Payment Management Solutions
CPOE- Computerized provider order entry	CM -Claims Management
LIS-Laboratory Information System	Others
PIS - Pharmacy Information System	

Figure 12 represents the market segment and the various models, applications and users of the cloud healthcare segment.APAC is said to emerge as the promising growth region in Healthcare Cloud Computing Market owing to increase in investments, infrastructural readiness and robust regulatory compliance's [19].The various drivers that are involved in the growth of the cloud computing market across the healthcare segment is presented in the form of a graph and accrue to favourable regulatory compliance's, technology upgrades and enhancement in digital literacy. The drivers that impede the growth of this market are primarily the data safety and security issues associated as reported in Fig. 13.

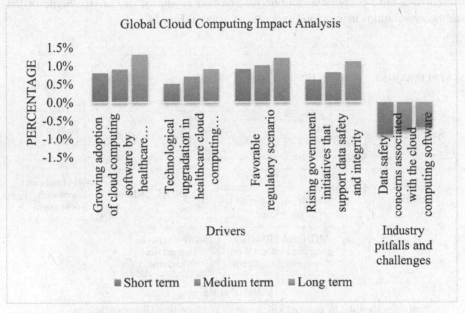

Fig. 13. Catalysts and challengers of global cloud healthcare market

4.1.2.2 Cloud Computing Market in Malaysian Healthcare Sector - A Regional Preview
In the segment of healthcare, a number of institutions in Malaysia have moved to the cloud, working in close collaboration with the Ministry of Health, Director General of Health and the Personal Data Protection Commissioner with staunch support from Microsoft. Malaysia's progress in healthcare is driven by the government initially pushing in US$ 5.2 bn in 2016 that was approximately 10% of the annual budget. Regulatory bodies like MSC Malaysia Cloud Initiative (MMCI) and the Malaysia Digital Economy Corporation (MDEC), advises the Malaysian government on the development of legislation, policies and the Personal Data Protection Act 2010 lays strict emphasis on the data retention and security policies. Some of the Healthcare providers of Malaysia that offer Hospital Information System on cloud and digital archival of radiology and pathology on cloud and health data warehouse and eklinik provisions are listed.

A survey of 135 public hospitals and 9 special medical institutions in Malaysia showcase that the most popular cloud healthcare application offering is the eHRMIS, MyCPD, and MyHealth, whereas Malaysian Pharmacies have opted e-store as their cloud solution and for the ambulatory units emasa is being used by the majority of the healthcare providers. For around 45 plus private hospitals in Malaysia,the popular cloud base solution offering used are the patient's appointment and enquiry system either offered by a third party or as an internal solution as shown in Fig. 16. Some of the healthcare solutions using cloud support are listed in Fig. 14.

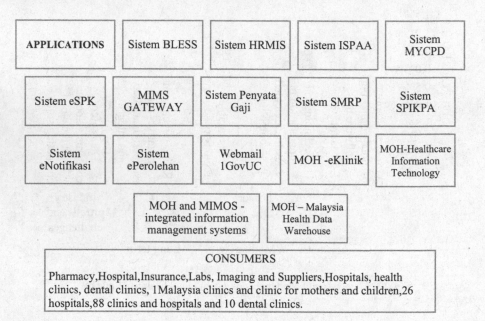

Fig. 14. Healthcare applications cloud

Around 45% of the Government Hospitals in Malaysia utilize cloud services and 70% of the Private Hospitals use applications hosted on cloud as in Fig. 14. Malaysia has envisaged in recent times many consumer health portals and tele-health solutions and data aggregation platforms that are being built to boost health systems.Several app based initiatives with respect to tourist health, employee wellness benefit and assisting the Human Resource (HR) department by managing complexities of employee healthcare plans to Doctors on Call and Medical Home Care Initiative have taken centre stage in Malaysia (Fig. 15).

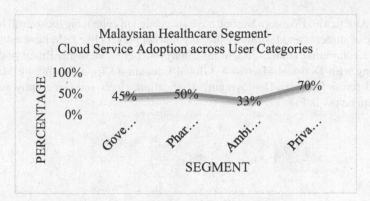

Fig. 15. Cloud service adoption-user categories

4.1.2.3 Cloud Computing Market in the Education Sector - A Global Preview

With the diffusion of online learning resources and adoption of blended initiatives by educational institutions across the globe it is anticipated that there is an emerging demand for Cloud Computing in Education Sector. Continued emphasis on the implementation of experimental and project based learning in the area of science, technology, engineering mathematics and management along with the recent advancements in artificial intelligence, virtual reality and big data has set the pace for cloud adoption.

Many educational institutes have resorted to the use of cloud computing supported through apps to keep tracks of student's performance and identifying their areas of weaknesses (Fig. 16).

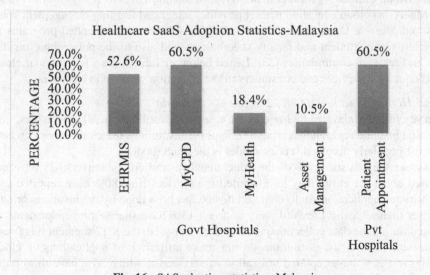

Fig. 16. SAS adoption statistics- Malaysia

North America and Western Europe have succeeded in enhancing the overall learning experience of students and faculties while big market players like IBM have established their cloud computing centres in China, Malaysia, India, Vietnam, Brazil and South Korea along with Dell and Microsoft. Global Education Cloud Computing Market is segmented across Services, Deployment, Application, Users and Geography as shown in the figure below in Fig. 17.

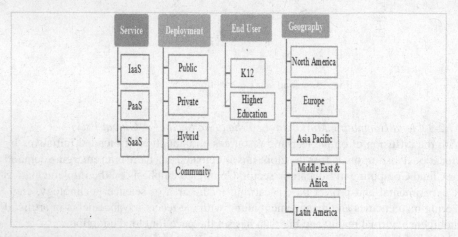

Fig. 17. Global education market segment

4.1.2.4 Cloud Computing Market in Malaysian Education Sector- A Regional Preview

The Malaysian cloud education market provides integrated learning management solutions and Massive Open Online Course (MOOC) offerings and blended programs to many University student and faculty stakeholders and also to the polytechnic institutions and research communities [23]. Listed below in Table 6 are a few popular cloud based education services and consumers in the education segment in Malaysia.

4.1.2.5 Higher Education and K 12 Segment - Malaysia

A survey of the Malaysian higher education sector constituting of 60 universities, colleges and institutes exhibit that around 67% of the institutions use cloud services where the most popularly accepted service model is the SaaS model.

As per the statistics, 5% of the educational institutions or universities have not reported about the entity they have availed the services from.80% have reported that platform and application configured and hosted has been done by the institutes or universities themselves or the third party and rest 13% have shared the vendor name as collated during the data collection process. With respect to the K 12 segment it is found that around 70% of the institutions do not make utilization of applications on cloud and the rest 30% have reported utilization of services of which 35% have third party assistance and support with respect to utilization of software as a service. The graph for the same is represented in Fig. 18.

Table 6. Cloud Applications for Education sector in Malaysia

Cloud education application/Product	Consumers	Cloud solution
Deskera Integrated Education Cloud/Learning Management System (LMS)	Universities/Institutions	Student, Faculty and Examination Management, Course Management, Learning Content Management, Institutional Finance Management, Social Media, Campus Management, Document Management, Virtual Class Room, Learning Portfolio Management
Open Learning	Taylor's University and all public universities, polytechnics and about a dozen private universities (700000 users)	MOOCs blended learning and online degree programmes
Curriculum Information Document Online System or the CIDOS	Malaysian Polytechnic	Competency Standards, Curriculum Inventory, Curriculum Evaluation, Curriculum Review Committee, Teaching & Learning Repository, Learning Management System
MOHE - Centre for Mobile Cloud Computing Research (C4MCCR)	Research Community	Medium between researchers and public, broadcasting information related to the laboratory activities and achievements

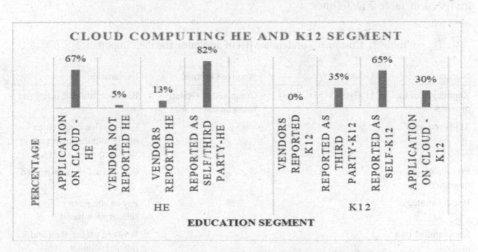

Fig. 18. Cloud computing trend in HE and K12 segment-Malaysia

4.1.2.6 Cloud Service Utilization Using Conjoint Analysis–Malaysian Healthcare and Education Segment

The choice of the cloud software adopted is typically based on the service attribute preference that can be concluded using the conjoint analysis. Conjoint analysis is an approach often used by academics, intended at scrutinizing consumers trade-offs among competitive products [15]. Some researchers have used an approach that identified that the most influencing cloud computing service characteristics, constituted of consistent service availability and lock-in prevention [22]. [24] has analyzed that consumers have never mentioned cost savings as a major consideration. The main hindrance to the cloud computing adoption is the attribute, information security, as investigated by [22, 25]. Also [26] cites that there is a shift in the focus of customers to service based considerations while using cloud computing services.

In the survey performed by [27], the researchers found using choice based conjoint analysis several attributes as follows: providers' reputation, required skills, migration process, pricing and tariff, cost compared to internal solution and consumer support, in which the first attribute was marked to have the highest relative importance. This paper has taken into consideration a multi-service provider's environment that maximizes consumer's utility based on his buying decisions. It uses a mathematical model to compute the consumer's preferences across the list of software as a service solutions, along the healthcare and education vertical. The data of the service provider attributes has been collected from the secondary sources. Also reports of consulting companies, case studies and white papers have been considered.

In order to estimate consumers' preferences, a choice based conjoint analysis, first introduced by [28] was taken into consideration. Service attributes and attribute levels were presented by [27] through a rigorous research in context of cloud services. A list of 18 attributes were identified by [21, 29] and there on they were reduced to 6 attributes list as detailed in the table below [24]. The list of attributes with the weighted average are listed in Table 7 as follows:

Table 7. Customer's preference list of attributes and their importance

Attribute	Grade	Weighted average	Explanation
Reputation of the provider	(1–3) or (1–5) (Likert)	Computed as (Relative Importance* Grade)	Based on attitude, belief and trust
Skills required			Easy service usage either due to training undergone or skills acquired
Process of migration			Customized or standardized
Pricing strategy			Pay as you go or subscription based
Comparative cost analysis			Between off the shelf and internal solution
Consumer support			Availability of FAQ's emails forum's etc.

The formulas used to calculate the relative importance of each is given as follows:

$$r_i(\text{best}) = \text{Ln}(B(i)/1 - B(i)))$$

$$r_i(\text{worst}) = r_i(w) = -\text{Ln}(W(i)/(1 - W(i)))$$

$$r_i(\text{average}) = \text{Average}(r_i(b), r_i(w))$$

$$r_i(\text{scaled}) = . r_i(i) - \text{Average}(r_i)$$ \hfill (2)

$$r_i(\text{transssformed}) = \text{Exp}((\text{Scaled})(i))$$

$$\text{Relative Importance} = . r_i(\text{transformed})(i)/\sum_{i=0}^{n} r_i/(\text{transformed})$$

From the previous discussion in this paper it is established that along the healthcare domain two SaaS repositories (HRMIS and MYCPD) by Ministry of Health-Malaysia have been widely used across the Government Hospitals accounting to 50% to 60% of the total utilization by the consumers. Whereas for the private units, the cloud based patient appointment, inquiry modules are the frequently used software's that alone surpasses 60% of the utilization. For the Education segment, the cloud service usage is more for the Higher Education Providers than the K-12 segment and the overall scenario validates the moderately paced growth in the adoption of cloud services by the nation. Multinomial Logit choice model (MNL) was considered for prediction of choice decision behaviour and utilities were identified using Max Likelihood and the relative importance deduced with high value of likelihood ratio [29].The Multinomial Logit Choice Model for prediction with variance in utilities of the services rendered is given as follows:

$$\frac{e^{u1}}{e^{u1} + e^{u2} + \ldots + e^{un}}$$ \hfill (3)

where u1, u2.unaretheutilityvalues.

The utility values are calculated using the formula as follows:

$$\text{Total parts worth} = p_1 + p_2 + p_3 + \ldots + p_n$$

$$\text{and Average Utility value} \frac{p_1 + p_2 + p_3 + \ldots + p_n}{n}$$ \hfill (4)

$$\text{Total Utility Computed/Conjoined Analysis Score} = p_1*r_{i1} + p_2*r_{i2} + p_3*r_{i3} + p_4*r_{i4}$$ \hfill (5)

Considering the consumer preference for the two Healthcare cloud service solutions, from the same service provider operating two different services, the mathematical model utilized to arrive at the necessary conclusion using secondary data source is presented in Table 8. Here the researchers consider customers preference rankings across attributes. The consumers have rated the above six attributes for a designated service parameter in the range of 1 to 3.The utility value is calculated using part worth value as marked by the consumer multiplied by the relative importance assigned in the form of the weight and adding all the values to find the final utility value as represented in Table 9. It is seen that for the Healthcare and the Education segment, the total utilization of the cloud based applications in context of users preferences are as follows: MyCPD (Healthcare) and Open Learning MOOC(Education).Then the average and the standard deviation recorded in evaluating the utilization of the said application establishes the fact that users had experienced maintaining of similar performance standards across all the attributes chosen, hence the deviation is minimal for those two applications having the highest utility value as represented in Table 10.

Table 8. Consumer Preference Ranking

SAAS	Service Name (Healthcare and Education)	Service Provider1 HRMIS, Ministry of Health (MOH),Malaysia(Utility z)	Service Provider1 MYCPDMOH,Malaysia (Utility)	Service Provider2 CIDOS-MOHE	Service Provider3 Open Learning
	Operational Continuity and Consistency	133322	232323	233312	332121
	Communication System-Mail,Msg,Alert	221213	323333	221211	121121
	Search Facility	132131	313332	232111	313332
	Content, Document Management	123123	313233	223121	313233
	Application Functionality	321123	333122	121121	333122
PAAS	Operating System(OS)	231123	321331	231121	321331
	Memory	211333	123321	211313	123321
	Storage	132321	331233	132321	331233
	Technical Support	3321124	231331	132112	231331
IAAS	Backup and Recovery	123313	321213	122313	321213
	Database	211231	311333	121233	311333
	Hard disk space	232321	333332	232311	333332

Further the Malaysian Healthcare consumer segment, generating, 45.51 utility units has a median of 7.57 whereas for the Education Services consumer, the application MOHE-OPEN LEARNING MOQC has a median value of 5.13.

Table 9. Conjoint Analysis Score

Healthcare Cloud Service	
Total utility Service Provider (SP1) MOH-MALAYSIA –HRMIS	= [.26(21) + 0.7(28) + .21(22) + .17(24) + .16(24) + .13(26)]
	= [5.46] + [18.2] + [4.62] + [4.08] + [3.84] + [3.38] = 39.58
Total utility Service Provider (SP1) MOH-MALAYSIA-MYCPD	= [.26(32) + 0.7(26) + .21(25) + .17(31) + .16(31) + .13(27)]
	= [8.32] + [18.2] + [5.25] + [5.27] + [4.96] + [3.51] = 45.51
Education	
Total utility Service Provider (SP1) MOHE-MALAYSIA-CIDOS	= [.26(19) + 0.7(29) + .21(21) + .17(22) + .16(18) + .13(20)]
	= [4.94] + [20.3] + [4.41] + [3.74] + [2.88] + [2.6] = 38.87
Total utility Service Provider (SP1) MOHE-MALAYSIA-OPEN LEARNINGMOOC	= [.26(31) + 0.7(26) + .21(23) + .17(27) + .16(30) + .13(23)]
	= [8.06] + [18.2] + [4.83] + [4.59] + [4.8] + [2.99] = 43.47

Table 10. Statistical evaluation of the utility values

Segment	W1 T1	W2 T2	W3 T3	W4 T4	W5 T5	W6 T6	Mean	Median	Std Deviation
Healthcare	5.46	18.2	4.62	4.08	3.84	3.38	6.28	4.08	5.7
	8.32	18.2	5.25	5.27	4.96	3.51	5.23	7.57	4.96
Education	4.94	20.3	4.41	3.74	2.88	2.6	4.08	6.48	6.23
	8.06	18.2	4.83	4.59	4.80	2.9	4.815	7.23	5.13

For both, the utility range spanning four out of six attributes is less than the other two applications which provides the justification of end users impression or level of satisfaction for the selected application categories also represented in Fig. 18. Performing a conjoint analysis and deducing inferences from the analysis helped in identification of the utility of each parameter and its contribution towards the success of the application. Further inclusion of standardized Electronic and Mobile Health records will necessitate the exploration of solutions which are compliant with the standard operating procedures and utility of parameters assessed accordingly [30] (Fig. 19).

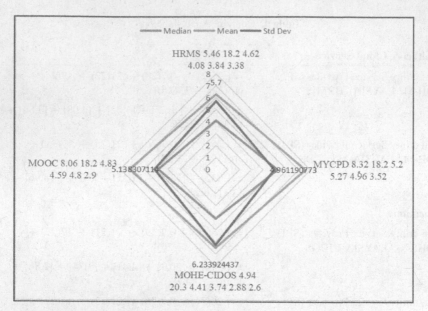

Fig. 19. Radar chart of application preferences (Statistical Analysis)

5   Conclusion and Future Work

The paper sets the base for post-doctoral research study undertaken in the vertical of cloud service adoption across Healthcare and Education segment of Malaysia to investigate the factors that contribute significantly to cloud technology adoption for ensuring greater streamlining, increased productivity in terms of effort expectancy, performance expectancy and social influence. Subsequent to the above, to establish through the conducted research. That adoption would result in sustainability of the organizations under these segments and take a shift from Capital Expenditure (CAPEX) to Operating Expenditure (OPEX) model aiming at reduces cost and effort expended. A summary of the research undertaken concludes that cloud service adoption is on the rise in both the urban, rural segment of Malaysia, with more government and semi government healthcare agencies, considering adoption based on user's preferences. While the big entities in the healthcare and education segment are considering adoption to build integrated platforms to serve the consumers and make available critical care to masses with greater flexibility and ease, the smaller business units are considering adoption for greater cost control and marketability.

References

1. Armbrust, M., et al.: Above the clouds: a Berkeley view a Berkeley view of cloud computing [technical report]. http://www.eecs.berkeley.edu/Pubs/TechRpts/2009/EECS-2009-28.html. Department of Electrical Engineering and Computer Sciences, University of California Berkeley, Technical Report No. UCB/EECS-2009-28 (2009)

2. 2018 BSA global cloud computing Scorecard. https://cloudscorecard. http://bsa.org/2018/pdf/country_reports/2018_Country_Report_Malaysia.pdf. Accessed 2021
3. Chen, D., Ma, M., Lv, Q.: A Federation model for education under hybrid cloud computing. Int. Confererence on Future Computers in Education **23–24**, 340–343 (2012)
4. Fern'ndez, G., de la Torre-Díez, I., Rodrigues, J.: Analysis of the cloud computing paradigm on mobile health records systems. In: Sixth International Conference on Innovative Mobile and Internet Services in Ubiquitous Computing, pp. 927–932 (2012)
5. Chen, Z., Han, F., Cao, J., Jiang, X., Chen, S.:Cloud computing based forensic analysis for collaborative network security management system. Tsinghua Sci. Technol.**18**(1), 40–50 (2013)
6. Forrester, J.S.: Cloud predictions for 2014:cloud joins the IT portfolio. cloud_computing_predictions_for_2014_cloud_joins_the_formal_it_portfolio (2004). http://blogs.forrester.com/james_staten/13-12-04. Accessed 18 Oct 2021
7. Deka, G.Ch., et al.: ICT's Role in e-governance in India and Malaysia: a review. J. Next Gene. Inf. Technol. **3**(2012)
8. Dan, M.A., Christoff, W.: Customer heterogeneity and tariff biases in cloud computing. In: ICIS 2010 Proceedings-Thirty First International Conference on Information Systems. Koehler, Philip and Anandasivam, Arun and ma, p. 106 (2010)
9. Index methodology: Data normalization, aggregation, and index construction,Source.https://giwps.georgetown.edu/wpcontent/uploads/2017/10/Appendix-1.pdf. Accessed 15 Oct 2021
10. Dominic, P., Ratnam, K.: Factor analysis: an applied approach towards the adoption of cloud computing to enhance the healthcare services in Malaysia. In: Information System International Conference (ISICO), pp. 79–84 (2013). https://doi.org/10.1038/sj.bjp.0707419
11. Healthcare cloud computing market size, global market insights.:https://www.gminsights.com/industry-analysis/healthcare-cloud-computing-market. Accessed 15 Oct 2021
12. MAMPU Laporan Pembentangan Dapatan dan Analisis Kajian Impak.: http://www.1govuc.gov.my/doc/Slaid%20Kajian%20Impak%20Tech%20Talk.pdf2018. Accessed 17 Oct 2021
13. Healthcare Cloud Computing Market:ESTIMATES and TREND ANALYSIS from 2014 to 2026,Grand View Research Inc., United States https://www.grandviewresearch.com/industry/healthcare. Accessed 18 Oct 2021
14. Green, P.E., Krieger, A.M., Wind, Y.J.:Thirty years of conjoint analysis: reflections and prospects. Interfaces **31**(3-suppl.), 56–73 (2001)
15. Hollobaugh.: Cloud Computing Trends Report [technical report:2009]. Hosting.com. http://www.hosting.com. Accessed 15 Oct 2021
16. Iyer, E.K., et al.: Sectorial adoption analysis of cloud computing by examining the dissatis-fier landscape. Electr. J. Inf. Syst. Eval. **16**(3), 211–219 published by Academic Publishing Limited, ISSN:1566-6379 (2013). http://www.ejise.com
17. Koehler, P., Anandasivam, A., Ma, D.: Cloud services from a consumer perspective. In: Proceedings of the 16th Americas Conference on Information Systems (AMCIS 2010), Lima, Peru, August 12–15. Research Collection School of Information Systems, pp. 1–11 (2011). https://ink.library.smu.edu.sg/sis_research/1313
18. Kuo, M.H., Kushniruk, A., Borycki, E.: Can cloud computing benefit health services? -A SWOT analysis. Stud. Health Technol. Inform. **169**, 379–383 (2011)
19. Kuziemsky, C., Peyton, L., Weber, J., Topalogou, T., Keshavjee, K.: 3rd Annual Work-shop on Interoperability and Smart Interactions in Healthcare (ISIH). In: Proceedings of the Conference of the Center for Advanced Studies on Collaborative Research, pp.351–352 (2011)
20. Louviere,J.J., Woodworth, G.: Design and analysis of simulated consumer choice or allocation experiments: an approach based on aggregate data. J. Mark. Res. **20**(4), 350–367 (1983). https://doi.org/10.1177/002224378302000403

21. Cloud computing in education market, global forecast to 2021, markets and market TC 2598, by service model (SaaS, PaaS, and IaaS),deployment model (private cloud, public cloud, hybrid cloud, community cloud), user type (K-12andhigher education) and region. https://www.marketsandmarkets.com/Market-Reports/cloud-computing-educat ion-market-17863862.html. Accessed 17 Oct 2021

22. Masrom, M., Rahimli, A., WNBinti, W.Z., Aljunid, S.M.: Understanding the problems and benefits of using cloud computing in Malaysia Healthcare Sector. Int. J. Adv. Comp. Eng. Netw. **4**(1), 58–62 (2016)

23. Ratnam, K.A., Dominic, P.D.D., Ramayah, T.: A structural equation modeling approach for the adoption of cloud computing to enhance the Malaysian healthcare sector. J. Med. Syst. **38**(8), 1–14 (2014). https://doi.org/10.1007/s10916-014-0082-5

24. Sheelvant, R.:10 Things to know about cloud computing strategy. IT Strategy (2009)

25. Weiber, R., Mühlhaus, D.: Auswahl von Eigenschaften und Ausprägungen bei der Conjoint Analyse. In: Baier, D., Brusch, M. (eds.) Conjoint Analyse, pp. 43–58. Springer, Heidelberg (2009). https://doi.org/10.1007/978-3-642-00754-5_3

26. Weintraub, E., Cohen, Y.: Optimizing user's utility from cloud computing services in a net-worked environment. Int. J. Adv. Comput. Sci. Appl. 6(10), 153 (2019). https://doi.org/ 10.14569/IJACSA.2015.061021. World Telecommunication/ICT Indicators Database (23rd ed/December 2019)

27. Sahharon, H., Omar, S.Z., Bolong, J., Mohamed, Sh. H.A., Lawrence, D.J.: Potential benefits of the wireless village programme in Malaysia for rural communities. J. Appl. Sci. **14**(24), 3638–3645 (2014). https://doi.org/10.3923/jas.2014.3638.3645

28. Venters, W., Whitley, E.A.: A critical review of cloud computing: researching desires and realities. J. Inf. Technol. **27**(3), 179–197 (2012)

29. Zhang, Q., Cheng, L., Boutaba, R.: Cloud computing: state-of-the-art and research challenges. J. Internet Serv. Appl. **1**(1), 7–18 (2010). https://doi.org/10.1007/s13174-010-0007-6

30. Fern'ndez, G., de la Torre-Díez, I., Rodrigues, J.: Analysis of the cloud computing paradigm on mobile health records systems. In: Sixth International Conference on Innovative Mobile and Internet Services in Ubiquitous Computing, pp. 927–932 (2012). A.List of websites referenced with respect to Malaysian Healthcare sector – 472. B. List of websites referenced with respect to Malaysian Education sector - 501

Generating Weakly Chordal Graphs
from Arbitrary Graphs

Sudiksha Khanduja[1], Aayushi Srivastava[1], Md Zamilur Rahman[1,2],
and Asish Mukhopadhyay[1(✉)]

[1] University of Windsor, Windsor, ON N9C1S7, Canada
{khanduj,sriva115,rahma11u,asishm}@uwindsor.ca
[2] Algoma University, Sault Ste. Marie, ON P6A2G4, Canada
zamilur.rahman@algomau.ca

Abstract. Algorithms for generating graphs that belong to a particular class are useful for providing test cases and counter-examples to refute conjectures about these graphs. This is true, in particular, for weakly chordal graphs. A graph G is weakly chordal if neither G nor its complement contains a chordless cycle of size greater than four. In an earlier paper, we proposed a separator-based scheme for generating weakly chordal graphs. In this paper, we propose a scheme to solve this open problem: generate a weakly chordal graph from a randomly generated input graph, G, adding as few edges as possible [2], unless the graph is already weakly chordal.

Keywords: Weakly chordal graph · Weak triangulation · Minimum triangulation

1 Introduction

Problem Definition: A graph $G = (V, E)$ is said to be weakly chordal if neither G nor its complement, $\overline{G}$, has an induced chordless cycle on five or more vertices (a hole). Figure 1 shows an example of a weakly chordal graph, G, and its complement, $\overline{G}$.

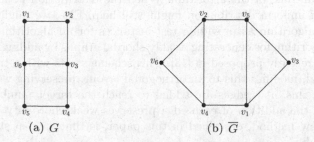

(a) G (b) $\overline{G}$

Fig. 1. (a) A weakly chordal graph; (b) its complement [13]

© Springer-Verlag GmbH Germany, part of Springer Nature 2022
M. L. Gavrilova and C. J. K. Tan (Eds.): Trans. on Comput. Sci. XXXIX, LNCS 13460, pp. 71–84, 2022.
https://doi.org/10.1007/978-3-662-66491-9_4

In [7], Hayward introduced weakly chordal graphs as a generalization of chordal graphs, and showed that these form a subclass of the perfect graphs. An alternate definition in terms of G alone is that it does not contain a hole or an antihole (the complement of a hole). Berry et al. [2] gave a very different and interesting definition of a weakly chordal graph as one in wich every edge is LB-simplicial (see Sect. 2). They also proposed the open problem of generating a weakly chordal graph from an arbitrary graph. A solution to this problem is the subject of this paper. Bouchitté and Todinca [3] showed that the minimum fill-in problem for weakly chordal graphs is polynomial-time solvable, thereby implying a method for generating chordal graphs from weakly chordal graphs, adding a minimum number of edges (called fill-edges, here and after). In this paper, the tables are turned. Our method shows how to generate a weakly chordal graphs from arbitrary graphs via chordal graphs, deleting as many fill-edges as possible.

Background and Motivation: Early work on graph generation focused on creating catalogs of graphs of small sizes. Cameron et al. [4], for instance, published a catalog of all graphs on 10 vertices. The underlying motive was that such repositories were useful for providing counterexamples to old conjectures and coming up with new ones. Subsequent focus shifted to generating graphs of arbitrary size, labeled and unlabeled, uniformly at random. As such a generation method, involved solving a counting problem, research was focused to classes of graphs for which the counting problem could be solved and yielded polynomial time generation algorithms. Among these were graphs with prescribed degree sequence, regular graphs, special classes of graphs such as outerplanar graphs, maximal planar graphs. See [15] for a survey of work prior to 1990.

As stated in [13], there are many situations where we would like to generate instances of these to test algorithms for weakly chordal graphs. For instance, in [12] the authors generate all linear layouts of weakly chordal graphs. A generation mechanism can be used to obtain test instances for this algorithm. It can do the same for optimization algorithms, like finding a maximum clique, maximum stable set, minimum clique cover, minimum coloring, for both weighted and unweighted versions, for weakly chordal graphs proposed in [8] and their improved versions in [9,14].

If the input instances for a given algorithm are from a uniform distribution, a uniform random generation provides test instances to obtain an estimate of the average run-time of the algorithm. When the distribution is unknown, the assumption of uniform distribution might still help. Else, we might look upon a generation algorithm as providing test-instances for an algorithm. With this motive, an algorithm for generating weakly chordal graphs by adding edges incrementally was recently proposed in [13]. This scheme starts with a tree, which is weakly chordal, modifies this to an orthogonal layout, preserving weak chordality. Following this, new edges are added to reach the target number of edges, ensuring that the addition of each edge preserves weak chordality. The advantage of the new method, described in this paper, is that we can start with an arbitrary graph.

Contents: The next section of the paper contains some common graph terminology, used subsequently. The following section contains details of our algorithms, beginning with a brief overview. In the concluding section, we summarize the salient aspects of the paper and suggest directions for further work.

2   Preliminaries

We will assume that G is a graph on n vertices and m edges, that is, $|V| = n$ and $|E| = m$. The *open neighborhood* $N(v)$ of a vertex v is the subset of vertices $\{u \in V \mid (u, v) \in E\}$ of V. The *closed neighborhood* $N[v]$ of v is the set $N(v) \cup \{v\}$. The *degree* $\deg(v)$ of a vertex v is equal to $|N(v)|$. If $S \subset V$, then the neighborhood of S is $N(S) = \bigcup_{x \in V} N(x) - S$. Clearly, the closed neighborhood of S is $N[S] = N(S) \cup S$. A vertex v of G is *simplicial* if the induced subgraph on $N(v)$ is complete (alternately, a *clique*). A *path* in a graph G is a sequence of vertices connected by edges. We use $P_k (k \geq 3)$ to denote a chordless path, spanning k vertices of G. For instance, a path on 3 vertices is termed as a P_3 and, similarly, a path on 4 vertices is termed as a P_4. If a path starts and ends in the same vertex, the path is a cycle denoted by C_k, where k is the length of the cycle. A *chord* of a cycle is an edge between two non-consecutive vertices in the cycle.

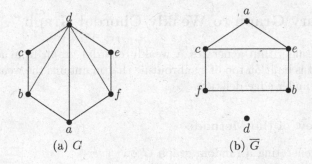

(a) G (b) $\overline{G}$

Fig. 2. Complement of a chordal graph with a chordless 4-cycle [13]

G is chordal if it has no induced chordless cycles of size four or more. However, as Fig. 2 shows, the complement of a chordal graph G can contain an induced chordless cycle of size four. The complement cannot contain a five cycle though, as the complement of a five cycle is also a five cycle (see Fig. 3). The above example makes it clear why chordal graphs are also weakly chordal.

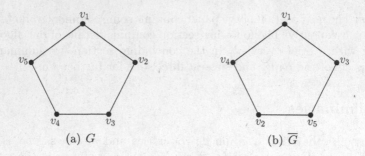

Fig. 3. Complement of a five cycle is also a five cycle

Let $S \subset V$; an edge $e \in G(V - S)$ is said to be S-saturating if all the vertices of each connected component S_j of $\overline{G}(S)$ (this is the induced graph on S in the complement of G) is visible from at least one of the end points of e Hayward [7].

An edge $e \in G$ is said to be *LB-simplicial* if one of the following two conditions holds [2]:

1. $e = \{u, v\}$ is S-saturating with respect to each minimal separator S contained in $N(\{u, v\})$;
2. $N(\{u, v\}) \cup \{u, v\} = V$.

3   Arbitrary Graph to Weakly Chordal Graph

Our proposed algorithm generates a weakly chordal graph from an arbitrary input graph. It is built on top of a subroutine that maintains the weak chordality of a graph G, under edge deletion.

3.1   Overview of the Method

We start by generating a random graph G on n vertices and m edges. Next, in a preprocessing step, we check if G is weakly chordal, using a recognition algorithm due to [2], based on the following characterization.

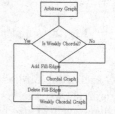

Theorem 1 [2]. *A graph is weakly chordal if and only if every edge is LB-simplicial.*

Fig. 4. Overview of process, pictorially

If G is weakly chordal, we stop. Otherwise, we proceed as follows. We first reduce G to a chordal graph H by introducing additional edges, called fill-edges, using the minimum degree vertex (mdv, for short) heuristic [6]. The mdv heuristic adds edges so that a minimum degree vertex in the current graph is simplicial. Each fill-edge is also entered into a queue, termed a fill-edge queue, FQ. These fill-edges are potential candidates for subsequent deletion from H. Since H is chordal, it is necessarily weakly chordal. We propose an algorithm for deleting edges from this weakly chordal graph to remove fill-edges, maintaining the weak chordality property. A fill-edge is deleted only if does not create a hole or an antihole in the resulting graph and we have developed criteria for detecting this. A fill-edge is removed from the front of the queue, which we then try to delete. If we do not succeed we put it at the back of the queue. We keep doing this until no more fill edges can be removed. Operationally, we implement this by defining a deletion round as one in which the fill-edge at the back of the queue is at the front. We stop when the size of the queue does not change over two successive deletion rounds. Figure 4 is a pictorial illustration of the flow of control.

3.2   Random Arbitrary Graph

To generate a random graph, we invoke an algorithm 'dense_gnm_random_ graph' by Keith M. Briggs. This algorithm, based on Knuth's Algorithm S (Selection sampling technique, see section 3.4.2 of [11]), takes the number of vertices, n and the number of edges, m, as input and produces a random graph. For a given n, we set m to a random value lying in the range between $n-1$ and $\frac{n(n-1)}{2}$. The output graph may be disconnected, in which case we connect the disjoint components, using additional edges.

3.3   Arbitrary Graph to Chordal Graph

By a process called triangulation (or fill-in), the arbitrary graph G is embedded into a chordal graph H by inserting fill-edges into G. Desirable triangulations are those in which a minimal or a minimum number of edges is added. A triangulation $H = (V, E \cup F)$ of $G = (V, E)$ is minimal if $(V, E \cup F')$ is non-chordal for every proper subset F' of F. Berry at al. [1] proposed a triangulation algorithm, known as LB-Triangulation, that provably adds a minimal number of fill-edges (see [1] for more details). In a minimum triangulation the number of edges added is the fewest possible. Since the minimum fill-in problem is NP-complete [16], several heuristics have been proposed in the literature. In this paper, we focus on the Minimum Degree Vertex (mdv) heuristic [6,10], discussed in some details in the next section.

The Minimum Degree Vertex Heuristic. Let $H = (V, E \cup F)$ be the graph obtained from $G = (V, E)$, by adding a set of fill-edges, F, obtained as follows. We first assign G to H and then prune from G all vertices of degree 1. From the remaining vertices of G we choose a vertex v of minimum degree (breaking ties arbitrarily) and turn the neighborhood $N(v)$ of v into a clique by adding edges as appropriate. These are fill-edges that we add to the set of edges of H, as well as to a queue called fill-queue, FQ (see Sect. 3.4), whose role will become apparent during the edge-deletion process. Finally, we remove from G, the vertex

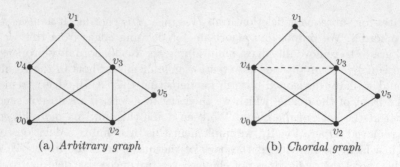

(a) *Arbitrary graph*   (b) *Chordal graph*

Fig. 5. Arbitrary graph to chordal graph

v and all the edges incident on it. We repeat this until G is empty. The graph H is now chordal. We illustrate this heuristic with an example.

The initial graph G is shown in Fig. 5a and the graph H with all fill-edges added is shown in Fig. 5b. In the initial graph G both v_1 and v_5 have minimum degree 2. We break the tie in favour of v_5. Since the induced subgraph on $N(v_5)$ is already a clique no fill-edges are added and G is set to $G - \{v_5\}$. In the reduced graph $G - \{v_5\}$, v_1 is of minimum degree and the induced graph on $N(v_1)$ is turned into a clique by adding $\{v_3, v_4\}$ as a fill- edge, which is also added to H. Since the reduced graph $G - \{v_1, v_5\}$ is a clique, we can pick the vertices v_0, v_2, v_3, v_4 in an arbitrary order to reduce G to an empty graph, without introducing any further fill-edges into H. Algorithm 1 is a formal description of this process.

What we have just described is a very basic version of the MDV heuristic. More sophisticated versions are available, designed with a view to keep the space complexity $O(n+m)$, as this is an essential requirement in the important application of this technique to the solution of sparse linear systems of equations. See [6,10] for further details.

3.4   Chordal Graph to Weakly Chordal Graph

Since the chordal graph H obtained from the previous stage is also weakly chordal, we apply an edge deletion algorithm to H that preserves weak chordality. The edges that are candidates for deletion are the fill-edges that have been added by the *mdv* heuristic. Each candidate edge is temporarily deleted from H, and we check if its deletion creates a hole or an antihole in H. If not, we delete this edge. The process is explained in details in the subsequent sections.

Fill-Edge Queue. As mentioned earlier, each edge added to convert an arbitrary input graph into a chordal graph is called a fill-edge. In order to delete as many fill-edges as possible, we maintain these edges in FQ. A fill-edge is removed from the front of this queue, which we then try to delete from H. If we do not succeed because a hole or antihole is created, we put it at the back of the queue

---

Algorithm 1. ArbitraryToChordal

Input: An arbitrary graph $G = (V, E)$

Output: Returns a chordal graph $H = (V, E \cup F)$ and fill-edge queue FQ

1: $H \leftarrow G$
2: Delete all vertices of degree 1 from G
3: Sort V in ascending order of degrees
4: Choose a vertex v of minimum degree
5: Turn $N(v)$ of v into a clique by adding edges, which are added to the edge set of
   H and to the fill-queue, FQ
6: Remove the vertex v from G and all the edges incident on it
7: Repeat steps 3 to 6 until G is empty

---

as it may become deletable at a later stage. We keep doing this until no more fill-edges can be removed from FQ.

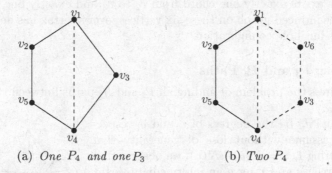

(a) *One P_4 and one P_3*   (b) *Two P_4*

Fig. 6. Detecting holes

Detecting Holes. To reiterate, a hole in a graph G is an induced chordless cycle on five or more vertices. Since, a graph is weakly chordal if it is (hole, antihole)-free [5], it is crucial to detect if any hole is formed by the deletion of an edge. For the class of weakly chordal graphs, since the biggest cycle allowed is of size four, a hole can be formed either by a combination of two P_4's or a by a combination of a P_3 and a P_4, as illustrated in Fig. 6.

To detect the formation of a hole in H, we pick an edge $e = \{u, v\}$ of H and temporarily delete it. To check if this deletion creates a hole in H, we find all P_3 and P_4 paths between u and v in H. A hole can be created in two distinct ways: (i) by a disjoint pair of P_4, with six distinct vertices between them such that there exist no chord joining an internal vertex on one P_4 to an internal vertex on the other; this we call a hole on two P_4s; (ii) by a disjoint pair of P_3 and P_4 between u and v, with five distinct vertices between them, such that there exist no chord joining an internal vertex on the P_4 to the internal vertex of the P_3; this we call a hole on a P_3 and a P_4.

Antiholes. An antihole in a graph is, by definition, the complement of a hole [5]. An antihole configuration in a weakly chordal graphs has the structure shown in Fig. 7. This is an induced graph on six distinct vertices each of which is of degree three.

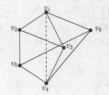

Fig. 7. An antihole

Detecting Antiholes. To detect an antihole configuration, we pick an edge $\{u, v\}$ and temporarily delete it from the graph. Next, we check if this creates an antihole configuration in the graph. For this, we find all chordless P_3 and P_4 paths between u and v. An antihole configuration is formed by a combination of two P_3 and one P_4 such that the induced graph on the six vertices that define these paths are uniformly of degree three, and there exists a chord from the internal vertex of each P_3 to one of the internal vertices on the P_4. For example, in Fig. 7, $\{v_1, v_2, v_5, v_4\}$ is a P_4, while $\{v_1, v_3, v_4\}$ and $\{v_1, v_6, v_4\}$ are two P_3 paths. There exists exactly one chord from v_2 to v_3 and exactly one from v_5 to v_6 and, in the induced graph on these six vertices, every vertex has degree three, making it an antihole configuration.

3.5   Finding P_3 and P_4 Paths

We now address the problem of finding all P_4 and P_3 paths between two vertices u and v in G.

Let d_u and d_v be the degrees of u and v, respectively, and assume without loss of generality that $d_u \geq d_v$. Setting $I_{u,v} = N(u) \cap N(v)$, we observe that all P_3's between u and v have an intermediate vertex in $I_{u,v}$ (see Fig. 8). Thus the number of such paths is $|I_{u,v}|$, which is bounded above by $min\{d(u), d(v)\} = d(v)$.

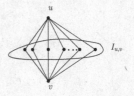

Fig. 8. Finding P_3 paths

As for P_4 paths, consider such a path from u to v (see Fig. 9): x is one of the at most d_u vertices adjacent to u and y is one of the at most d_v vertices adjacent to v. It can be seen from Fig. 10 that the number of such P_4 paths is bounded above by $O(d_u d_v)$, where d_u and d_v are the degrees of the vertices u and v respectively.

Fig. 9. A P_4-path from u to v

Let $X = N(u) - I_{u,v}$ and $Y = N(v) - I_{u,v}$. Edges such as xy shown in Fig. 9 are found by determining for each $x \in X$, its neighbors $y \in Y$. The computation of all P_4 paths between u and v, as seen in Fig. 10, is now straightforward, with time-complexity bounded above by $O(d_u d_v)$.

Proposed Algorithm. Our algorithm is built on top of a subroutine for deleting fill-edges from a weakly chord graph, maintaining its weak chordality property. In order to delete as many fill-edges as possible, a fill-edge $\{u, v\}$ is removed from the front of the fill-queue, which we then try to delete from H. If we do not succeed, we put it at the back of the queue. We keep doing this until no more fill-edges can be removed. Operationally, we implement this by defining a deletion round as one in which the edge at the back of the queue is at the front. One deletion round comprises of picking an edge from the start of the queue and deleting it from H. Now we check if the deletion of $\{u, v\}$ creates a hole or an antihole in H. If so, we do not delete the edge $\{u, v\}$ and add it back to the fill-queue. Otherwise, we delete the edge from H and also remove it from te fill-queue FQ. We stop when the size of FQ does not change over two successive deletion rounds.

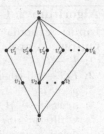

Fig. 10. Finding P_4 paths

Consider, for example, the random graph of Fig. 11 on 6 vertices and 8 edges. By inserting the edges $[\{v_2, v_4\}, \{v_1, v_4\}]$, we have a chordal graph, H. The added edges are put into the fill-edge queue. Maintain a temporary copy of the chordal graph H in T. The deletion algorithm begins by picking the first edge $\{v_2, v_4\}$ from the fill-edge queue and temporarily deletes it from graph T to check for hole and antihole configurations. Since deleting $\{v_2, v_4\}$ does not give rise to any hole or antihole configurations, $\{v_2, v_4\}$ is permanently deleted from the initial graph H, which is surely weakly chordal. The deletion algorithm picks the remaining edge $\{v_1, v_4\}$ from the fill-edge queue and temporarily deletes it from graph T and checks if this creates a hole or an antihole configuration. Since deleting $\{v_1, v_4\}$ gives rise to a hole configuration- a $P_4 = \{v_1, v_2, v_3, v_4\}$ with a $P_3 = \{v_1, v_5, v_4\}$ - $\{v_1, v_4\}$ is not permanently deleted from H. Since the queue is now empty, the graph G_w returned by the algorithm is weakly chordal with only the fill-edge $\{v_1, v_4\}$, added to the original graph G.

For another example, consider the random graph of Fig. 12a on 6 vertices and 9 edges. Adding the edges $[\{v_1, v_5\}, \{v_2, v_4\}, \{v_1, v_4\}]$ makes it a chordal graph H (see Fig. 12b). The added are put into the fill-edge queue. Maintain a temporary copy of the chordal graph H in T. The deletion algorithm begins by picking first edge $\{v_1, v_5\}$ in the fill-edge queue and temporarily deletes it from graph T to check if a hole or an antihole configuration is created. However, this does not happen so $\{v_1, v_5\}$ is permanently deleted from H, which is now surely a weakly chordal (see Fig. 12c). The fill-edge queue is updated and its contents are now: $[\{v_2, v_4\}, \{v_1, v_4\}]$. The deletion algorithm now picks the first edge in $\{v_2, v_4\}$ in the fill-edge queue and temporarily deletes it from graph T to check for hole and antihole configurations. Since deleting $\{v_2, v_4\}$ does not give rise to any hole or antihole configuration, $\{v_2, v_4\}$ is permanently deleted from starting graph H, which is now a weakly chordal graph. Now the updated fill-edge queue is $[\{v_1, v_4\}]$. The deletion algorithm now picks the first and only edge $\{v_1, v_4\}$ in the fill-edge queue and temporarily deletes it from graph T to

Algorithm 2. ChordalToWeaklyChordal

Input: A chordal graph $H = (V, E \cup F)$ with fill edge queue F
Output: A weakly chordal graph G_w

1: $T \leftarrow H$ ▷ Make a copy of H
2: $FQ \leftarrow$ fill-edges of H
3: $prevSize \leftarrow 0$
4: $newSize \leftarrow |FQ|$
5: **while** $(prevSize \neq newSize \ \&\& \ newSize \neq 0)$ **do**   ▷ Check size of FQ over two
   deletion rounds
6:     $prevSize \leftarrow newSize$
7:     **for** $(each \ edge\{u, v\} \ in fill\text{-}queue, \ FQ)$ **do**
8:         Delete edge $\{u, v\}$ from T
9:         **if** $(Hole \ or \ Antihole \ Detected)$ **then**
10:             Do not delete edge from graph H, add edge back to temporary graph T,
    and to the back of the queue FQ
11:         **else**
12:             Delete edge $\{u, v\}$ from graph H
13:         **end if**
14:     **end for**
15:     $newSize \leftarrow |FQ|$
16: **end while**
17: $G_w \leftarrow H$
18: **return** G_w

check for a hole or an antihole configuration. Since deleting $\{v_1, v_4\}$ gives rise to an antihole configuration on two P_3 paths $\{v_1, v_3, v_4\}, \{v_1, v_6, v_4\}$ and one $P_4 = \{v_1, v_2, v_5, v_4\}$ path, the edge $\{v_1, v_4\}$ is not permanently deleted from starting graph H. Since the queue is now empty, the graph G_w returned by the algorithm is weakly chordal with only the fill-edge $\{v_1, v_4\}$, added to the original graph G (see Fig. 12d).

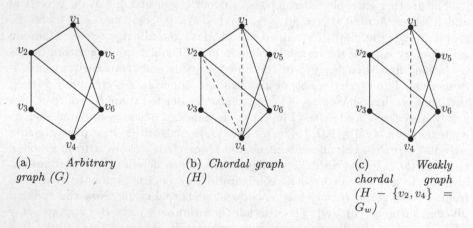

(a)   *Arbitrary graph (G)*

(b) *Chordal graph (H)*

(c)   *Weakly chordal graph* $(H - \{v_2, v_4\} = G_w)$

Fig. 11. Arbitrary graph to weakly chordal graph

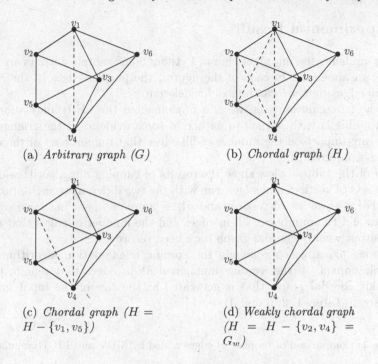

(a) *Arbitrary graph (G)*

(b) *Chordal graph (H)*

(c) *Chordal graph (H = H − {v₁, v₅})*

(d) *Weakly chordal graph (H = H − {v₂, v₄} = G_w)*

Fig. 12. Arbitrary graph to weakly chordal graph

3.6 Complexity

The *mdv* heuristic can be implemented in $O(n^2 m)$ time [10], while the time-complexity of the recognition algorithm based on LB-simpliciality is in $O(nm)$ [2].

To bound the query complexity of deleting an edge $\{u, v\}$ from the weakly chordal graph, the discussions of the previous sections show that we have to check if this creates a hole or an antihole.

The task of detecting a hole involves checking pairs of (P_4, P_4) and (P_4, P_3) paths for chords going between internal vertices of these paths. From the discussion of Subsect. 3.5, it follows that the time-complexity of this is in $O(d_u^2 d_v^2)$.

To detect an antihole we have to check for each triplet of paths (P_4, P_3, P_3) if the induced graph on all the path vertices form a configuration shown in Fig. 7. The number of such triplets is bounded above by $O(d_u d_v^3)$.

Thus an upper bound on the query complexity is $O(d_u^2 d_v^2)$, since we have assumed that $d_u \geq d_v$. The deletion of an edge takes constant time since we maintain an adjacency matrix data structure to represent G.

4   Experimental Results

We have implemented our algorithm in Python. Some sample outputs are shown below in an appendix. In each of the figures, the purples edges in the chordal graph are edges that are candidates for deletion.

For the triangulation step we also implemented the LB-triangulation algorithm [1]. This is with a view to gathering some evidence if the triangulation step has any impact on the number of fill-edges that remain part of the output weakly chordal graph.

Each of the tables below show the results of running our algorithm on four different sets of vertices and edges, run with the two different triangulation algorithms. For a given set of vertices and edges, an experiment has been repeated 5–6 times and the number of fill-in edges and the number of edges that remain in the output weakly chordal graph have been recorded.

It seems plausible to conclude the specific triangulation algorithm has a noticeable impact on the average number of fill-in edges that remain part of the weakly chordal graph that is generated as the size of the input graph G grows larger (Tables 1, 2, 3 and 4).

Table 1. Comparison of number of edges added by MDV and LB-Triangulation

Arbitrary graph {V, E}	Fill edges added by MDV	Final edges left in WCG	Fill edges added by LBT	Final edges left in WCG
{10, 30}	4	30	4	30
{10, 30}	4	31	6	31
{10, 30}	8	33	14	33
{10, 30}	4	30	4	30
{10, 30}	7	32	9	33
{10, 30}	3	30	5	30

Table 2. Comparison of number of edges added by MDV and LB-Triangulation

Arbitrary graph {V, E}	Fill edges added by MDV	Final edges left in WCG	Fill edges added by LBT	Final edges left in WCG
{15, 20}	6	23	5	22
{15, 20}	8	23	8	24
{15, 20}	4	21	4	21
{15, 20}	4	22	4	22
{15, 20}	14	27	11	26

Table 3. Comparison of number of edges added by MDV and LB-Triangulation

Arbitrary graph {V, E}	Fill edges added by MDV	Final edges left in WCG	Fill edges added by LBT	Final edges left in WCG
{50, 100}	123	193	207	238
{50, 100}	148	214	237	258
{50, 100}	98	170	211	246
{50, 100}	137	198	191	216
{50, 100}	117	184	229	246

Table 4. Comparison of number of edges added by MDV and LB-Triangulation

Arbitrary graph {V, E}	Fill edges added by MDV	Final edges left in WCG	Fill edges added by LBT	Final edges left in WCG
{100, 150}	211	311	506	526
{100, 150}	226	325	440	458
{100, 150}	256	368	482	504
{100, 150}	170	272	395	417
{100, 150}	208	303	433	454

5 Conclusion

We have proposed a simple method for generating a weakly chordal graph from an arbitrary graph. The proposed algorithm can also be used to generate weakly chordal graphs by deleting edges from input graphs that are known to be weakly chordal, such as complete graphs. Starting with complete graphs also helps in generating dense weakly chordal graphs. The algorithm we have proposed is contingent on first producing a minimum triangulation by the MDV heuristic. As a consequence, it is difficult to characterize the edges that are present in the final weakly chordal graph. An interesting and challenging problem is to design an algorithm for generating a weakly chordal graph directly from an arbitrary graph, akin to the LB-triangulation algorithm of Berry [1] for chordal graphs and establish that the weak triangulation adds a minimal number of edges.

Conflict of Interest. None.

References

1. Berry, A.: A wide-range efficient algorithm for minimal triangulation. In: Proceedings of the Tenth Annual ACM-SIAM Symposium on Discrete Algorithms, Baltimore, Maryland, USA, 17–19 January 1999, pp. 860–861 (1999)
2. Berry, A., Bordat, J.P., Heggernes, P.: Recognizing weakly triangulated graphs by edge separability. Nord. J. Comput. **7**(3), 164–177 (2000)
3. Bouchitté, V., Todinca, I.: Treewidth and minimum fill-in: grouping the minimal separators. SIAM J. Comput. **31**(1), 212–232 (2001)

4. Cameron, R.D., Colbourn, C.J., Read, R.C., Wormald, N.C.: Cataloguing the graphs on 10 vertices. J. Graph Theory **9**(4), 551–562 (1985)
5. Feghali, C., Fiala, J.: Reconfiguration graph for vertex colourings of weakly chordal graphs. CoRR, abs/1902.08071 (2019)
6. George, A., Liu, J.W.H.: The evolution of the minimum degree ordering algorithm. SIAM Rev. **31**(1), 1–19 (1989)
7. Hayward, R.B.: Weakly triangulated graphs. J. Comb. Theory Ser. B **39**(3), 200–208 (1985)
8. Hayward, R.B., Hoàng, C.T., Maffray, F.: Optimizing weakly triangulated graphs. Graphs Comb. **5**(1), 339–349 (1989). https://doi.org/10.1007/BF01788689
9. Hayward, R.B., Spinrad, J.P., Sritharan, R.: Improved algorithms for weakly chordal graphs. ACM Trans. Algorithms **3**(2), 14 (2007)
10. Heggernes, P., Eisenstat, S.C., Kumfert, G., Pothen, A.: The computational complexity of the minimum degree algorithm. Technical report (2001)
11. Knuth, D.E.: The Art of Computer Programming, Volume 2/Seminumerical Algorithms, 3rd edn. Addison-Wesley (1997)
12. Mukhopadhyay, A., Rao, S.V., Pardeshi, S., Gundlapalli, S.: Linear layouts of weakly triangulated graphs. Discrete Math. Algorithms Appl. **8**(3), 1–21 (2016)
13. Rahman, M.Z., Mukhopadhyay, A., Aneja, Y.P.: A separator-based method for generating weakly chordal graphs. Discrete Math. Algorithms Appl. **12**(4), 2050039:1–2050039:16 (2020)
14. Spinrad, J.P., Sritharan, R.: Algorithms for weakly triangulated graphs. Discrete Appl. Math. **59**(2), 181–191 (1995)
15. Tinhofer, G.: Generating graphs uniformly at random. In: Tinhofer, G., Mayr, E., Noltemeier, H., Syslo, M.M. (eds.) Computational Graph Theory. COMPUTING, vol. 7, pp. 235–255. Springer, Vienna (1990). https://doi.org/10.1007/978-3-7091-9076-0_12
16. Yannakakis, M.: Computing the minimum fill-in is np-complete. SIAM J. Algebraic Discrete Methods **2**(1), 77–79 (1981)

A Novel Machine Learning Framework for Covid-19 Image Classification with Bio-heuristic Optimization

Prathap Siddavaatam, Reza Sedaghat$^{(\boxtimes)}$, and Nahid Sahelgozin

OPRA-Labs, Ryerson University, 350 Victoria Street,
Toronto, ON M5B 2K3, Canada
{prathap.siddavaatam,rsedagha,nsahelgozin}@ee.ryerson.ca
https://www.ee.ryerson.ca/opr/

Abstract. Due to its rapidly advancing spread, the world is still reeling from COVID-19 (coronavirus 2019), which is categorized as a highly infectious disease. An early diagnosis is very critical in treating COVID-19 patients due to its lethal implications. However, the shortage of X-ray machines has resulted in life-threatening conditions and delays in diagnosis, increasing the number of deaths around the world. Therefore, in order to avoid such fatalities, COVID-19 has to be detected earlier and diagnosed faster using an intelligent computer-aided diagnosis system than with traditional screening programs. We present a novel framework for COVID-19 image categorization in this article that utilizes deep learning (DL) and bio-inspired optimization techniques. A bio-heuristic optimizer algorithm MoFAL is utilized as a feature selector to decrease the dimensionality of the image representation and increase the accuracy of the classification by ensuring that only the most essential selected features are used. Furthermore, the feature extraction is realized using the MobileNetV3 DL model. The experimental results deduced indicate that our proposed approach drastically improves performance in terms of classification accuracy and reduction in dimensions reflected during the period of feature extraction and its phases of selection. We propose that our COVID-Classifier can be deployed in conjunction with other tests for optimal allocation of hospital resources by rapid triage of non-COVID-19 cases.

Keywords: Feature extraction · Feature selection · Swarm optimization · Hybrid evolutionary optimizer · Machine learning · COVID-19 · Bio-heuristics · Machine learning · Artificial intelligence

1 Introduction

The severe acute respiratory syndrome coronavirus 2 (SARS-CoV-2) was first detected in Wuhan, China, at the end of year 2019. This disease, which is stated to be transmitted through the respiratory tract, was officially named COVID-19 by the World Health Organization on February 12, 2020. Thereafter by spreading

© Springer-Verlag GmbH Germany, part of Springer Nature 2022
M. L. Gavrilova and C. J. K. Tan (Eds.): Trans. on Comput. Sci. XXXIX, LNCS 13460, pp. 85–108, 2022.
https://doi.org/10.1007/978-3-662-66491-9_5

around the world through human-to-human transmission from 2019 until now, COVID-19 has been declared a global pandemic (Wang et al. 2020a; Wang et al. 2020c).

As a consequence of the contagious nature of COVID-19, it has rapidly spread to over 220 countries around the world with confirmed cases approaching 200 million and resulting in 4.3 million deaths worldwide. Statistically based on the number of deaths, this pandemic ranks second among all reported pandemics after the 1918 flu pandemic (Boberg-Fazlic et al. 2021). In fact, more than 40% of all confirmed COVID-19 cases and deaths is recorded in more populous countries such as USA, Brazil and India. Patients infected with COVID-19 may experience a variety of symptoms, including fever, dry cough, and respiratory tract infections. These symptoms can lead to pneumonia, respiratory distress, organ failure, and even death in extreme situations (Zhu et al. 2020). In the recent past there has been an uptick in AI packages used to diagnose and predict COVID19 (Oh et al. 2020; Roy et al. 2020; Wang et al. 2020b). Several methods have since been developed to determine and differentiate between COVID-19 and common flu infections utilizing chest radiography and computerized tomography (CT) scans (Dai et al. 2020).

X-ray imaging of the lungs has been extensively used to identify COVID-19 infection and is considered to be one of the most deterministic diagnosis tool (Kesim et al. 2019; Dai et al. 2020). In contrast, the traditional human-based detection technique, which solely relies on a technical expertise of a physician or a radiologist technical is typically inefficient or inaccurate leading to misdiagnosis and many a times delayed results in delayed diagnosis (Onder et al. 2021). Furthermore, such cases of misdiagnosis is more acute in remote areas where there is significant dearth of specialist medical personnel. Consequently, many advances in Artificial Intelligence (AI) have helped paved the way to build better tools like software expert systems, and instruments for improving diagnostics.

Moths are fancy insects, which are highly similar to the family of butterflies. Basically, there are over 160,000 various species of this insect in nature. They have two main milestones in their lifetime: larvae and adult. The larvae is converted to moth in cocoons. The most interesting fact about moths is their special navigation methods in night. They have been evolved to fly in night using the moon light. They utilized a mechanism called transverse orientation for navigation. In this method, a moth flies by maintaining a fixed angle with respect to the moon, a very effective mechanism for travelling long distances in a straight path [?]. Since the distance to moon is insurmountable for the moth, this mechanism guarantees flying in straight line.

A detailed relationship between CT imagery and COVID19 based chest pneumonia was established in (Demirci et al. 2021) and the results led to observing typical features in examined images of COVID19 cases, thereby allowing researchers to apply AI in the image processing of chest X-rays and CT of COVID-19 cases. With the onset of symptoms, CT imagery of the infected COVID-19 cases usually show bilateral and peripheral symptoms, greater total lung involvement, ground-glass opacities, "crazy-paving" patterns and the "reverse halo" signs as detailed in (Bernheim et al. 2020) by Mei et al. A convolutional neural network (CNN) based model was proposed in (Chaddad et al. 2021) to detect ground-glass opacities in the CT images of COVID-19 infected cases and subsequently a correlation between the severity of COVID-19 infection and patients' sex and age was determined in (Larici et al. 2021), by analyzing their X-ray images.

Machine learning (ML) based methods have shown unprecedented success in the reliable analysis of medical images (Khuzani et al. 2021) as result of their scalability, automation capability and portability (Ahmed et al. 2020). A typical approach for classification of images with highly similar features is to deploy ML-based image analysis. This analysis relies on the segmentation of image region of interest, identification of effective image features extracted from the segmented area in the spatial or frequency domain, and development of an optimal machine learning-based classification method to accurately assign image samples into target classes (Sun et al. 2020). Numerous ML-based methods have been deployed for the diagnosis of COVID-19 (Khuzani et al. 2021). ML based models like ResNet–18, ResNet–50, ResNet–50V2, Xception, AlexNet, LeNet-5, CoroNet are some of the prominent models used to detect COVID-19 infections based on CT or X-ray imagery (Kundu et al. 2021; Maior et al. 2021). Several hybrid approaches of machine learning and metaheuristic algorithms have been proposed recently (Canayaz 2021; Narin 2021) that outline feasibility of the merger resulting in sub-optimal computational costs. Recently, a deep neural network-integrated with metaheuristic optimizers to diagnose COVID-19 infections was proposed (Canayaz 2021) which uses a dataset with three classes of X-ray images: basic, pneumonia, and COVID-19, and the model was trained. In (Canayaz 2021), feature extraction was done using DL models like GoogleNet, VGG19, AlexNet, and ResNet–18 and best features were chosen using double metaheuristic algorithms such as Grey Wolf (Siddavaatam and Sedaghat 2019) and Particle Swarm optimizer culminating in support vector machine classification of features. Furthermore, a similar ResNet–18 classifier model was used for three classes: COVID-19, pneumonia, and normal – generating an accuracy of 92.49% in the application of the lung-lesion segmentation in CT images (Zhang et al. 2020).

Recently, there has been a surge in utilization of combined models of ML and metaheuristic-based classifications for effective feature extraction of chest X-ray and CT images (Sahlol et al. 2020; Elaziz et al. 2020; Yousri et al. 2021; Elaziz et al. 2020). Consequently, these models for COVID19 image classification have several drawbacks that impact accurate classification and prediction. Most of these drawbacks stem from dual impact of ineffective strategy employed for feature extraction and feature reduction. Therefore, in this article we hypothesize that chest X-ray and CT images can be reliably distinguished from other forms of pneumonia using an effective hybrid ML-Metaheuristic classifier, by employing reduction of dimensions to generate a model with an optimized set of features for differentiation. Furthermore, the approach taken in this article can be easily extrapolated in any future viral outbreak for the rapid classification of chest X-ray and CT images. We introduce a novel COVID-19 image classification technique that combines the advantages of an effective ML feature extractor to learn and extract relevant image representations using MobileNetV3 (Howard et al. 2019) and a novel bioheuristic optimizer MoFAL (Siddavaatam and Sedaghat 2021). The features are extracted from the tested images using MobileNetV3 followed by selection of features using the MoFAL optimizer. In our article (Siddavaatam and Sedaghat 2021), it was deduced that the MoFAL optimizer is capable of finding superior optimal designs for various applications that include diverse search spaces and select features more efficiently than many traditional evolutionary algorithms, and in addition, it avoids the occurrence of premature phenomena during its convergence process, and hence MoFAL has been chosen exclusively for abetting the feature extraction process for the model in this article.

The paper is organized as follows: Sect. 2 describes the proposed model architecture; Sect. 3 describes the simulation setup for the hybrid model followed by the results of the performance of the classifier compared to other state-of-the art architectures while Sect. 4 provides the conclusion of the paper.

2   Preliminaries

We propose an ensemble of two stages described in Fig. 1. First, during the initial stage, the fine tuning process of MobileNetV3 for feature extraction phase is utilized with the intent of extracting relevant image embeddings relying on a pre-trained model for different COVID-19 image datasets. Second, it is followed up by applying MoFAL optimizer during the next stage for feature selection procedure.

2.1   MobileNetV3

In the recent past, ML based convolutional neural network architectures are being deployed to tackle many real-world problems and scale their performance in terms of speed and size. A plethora of applications such as computer vision, robotics, etc. (Tran et al. 2019; Ji et al. 2020; Liu et al. 2021; Ignatov et al. 2021) utilize highly efficient convolutional neural networks (CNNs) applying contemporary depth-wise convolution structures like Visual Geometry Group (VGG19) (Simonyan and Zisserman 2015), DenseNet (Huang et al. 2017), EfficientNet (Ciga et al. 2021), NASNet (Zoph et al. 2018), MobileNets (Harjoseputro et al. 2020; Howard et al. 2019), MnasNet (Tan et al. 2019), and ShuffleNets (Zhang et al. 2018). Depth-wise convolutional kernels are shared across all input channels, increasing model efficiency and reducing computation cost since it involves a learnable parameter fed individually to each input channel separately from the training images to extract spatial information. A common drawback to such an approach is that the size of the depth-wise convolution kernel can be difficult to learn, thus increasing the complexity of the training process. Previous iterations MobileNetV1 and MobileNetV2 were improved to mitigate this specific drawback

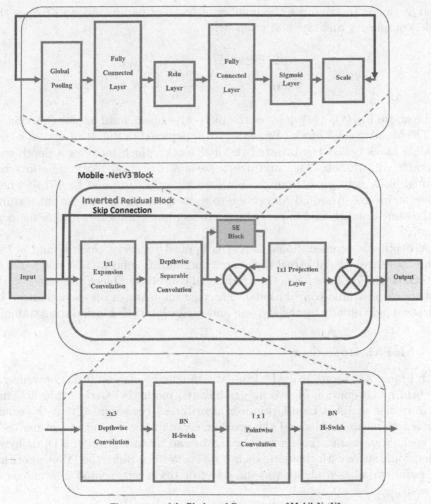

The structure of the Blocks and Components of MobileNetV3

Fig. 1. MobileNetV3 structure

of depth-wise convolution kernels and a new version called MobileNetV3 was introduced by Howard et al. (2019) that uses network architecture search (NAS) model. This particular NAS technique called NetAdapt algorithm was employed to search for the best kernel size and search the optimized MobileNet architecture to fulfill the low-resourced hardware platforms constrained by size, performance, and latency. Figure 2 illustrates the MobileNetV3 architecture derived from its previous versions MobileNetV1 and MobileNetV2. MobileNetV3 introduces a new activation function called hard-swish (`h-swish`). The `h-swish` is a type of activation function based on Swish (Ramachandran et al. 2017), but replaces the computationally expensive sigmoid (Ramachandran et al. 2017) with a piecewise linear analogue to minimize the number of training parameters and reduce the model complexity and size and given by:

$$\texttt{h-swish}(x) \;=\; z \,\cdot\, \gamma(z) \tag{1}$$

$$\text{where } \gamma(z) \;=\; \frac{\texttt{ReLU6}(z+3)}{6}$$

In the above Eq. (1), $\gamma(z)$ denotes the piece-wise linear hard analog function.

The MobileNetV3 block (Tan et al. 2019) depicted in Fig. 1 consists of a core building block called the inverted residual block, which includes a depth-wise separable convolution block and a squeeze-and-excitation block. The inverted residual block is inspired from the bottleneck blocks (Tan and Le 2019), where it uses an inverted residual connection to connect the input and output features on the same channels and improve the features representations with low memory usage.

A depth-wise convolutional kernel is applied to every channel and a 1×1 pointwise convolutional kernel with batch normalization layer (BN) and h-swish or ReLU activation functions. The traditional convolution block is altered by the depth-wise separable convolutional. The relevant features on each channel are addressed individually by the squeeze-and-excitation (SE) block during training.

2.2 MoFAL Optimizer

Moth Flame-Ant Lion (MoFAL) (Siddavaatam and Sedaghat 2021) optimizer is a hybrid amalgamation of two nature inspired methods based on Moth Flame Optimization and Ant Lion Optimizer algorithms. It is assumed that the candidate solutions are moths and the problem's variables are positions of moths in the space dimensions. Therefore, the moths can fly in 1-D, 2-D, 3-D, or hyper dimensional space with changing their position vectors. Since the MFO algorithm is a population-based algorithm, the set of moths is represented in a matrix as follows:

$$\mathbf{M} = \begin{bmatrix} M_{1,1} & M_{1,2} & M_{1,3} & \ldots & M_{1,d} \\ M_{2,1} & M_{2,2} & M_{2,3} & \ldots & M_{2,d} \\ \multicolumn{5}{c}{\ldots\ldots\ldots\ldots} \\ M_{n,1} & M_{n,2} & M_{n,3} & \ldots & M_{n,d} \end{bmatrix} \tag{2}$$

where n is the number of moths and d is the number of variables (dimension). Also, for all the moths, it is assumed that an array exists which holds the related fitness solutions of the moths and it is given by the following equation:

$$\mathbf{OM} = \begin{bmatrix} OM_1 \\ OM_2 \\ \vdots \\ \vdots \\ OM_n \end{bmatrix} \tag{3}$$

Data: $W_n = 50$ – Number of search agents, Iterations$_{MAX} = 500$ – total
number of iterations more than 0, $d = 30$ – Number of variables,
Random $\ell \in [y, 1], y$ linearly decreases from -1 to -2

Result: Optimal MoFAL flame value within given feature space of 70% image
dataset.

begin

    Initialize set of moths matrix – Eq. 2

    Calculate corresponding moth fitness value – Eq. 3

    Initialize position of ants – Eq. 8

    Evaluate fitness of each ants – Eq. 9

    Find best fitness antlion and assume it as Elite.

    Initialize $x := 0$

    repeat

        Update Flame$_{number}$ – Eq. 21

        if $x == 1$ **then**

            F := sort(M) – Eq. 4

            OF := sort(OM) – Eq. 5

        end

        else

            F := sort(M_{x-1}, M_x)

            OF := sort(M_{x-1}, M_x)

        end

        forall *moths/ants* $\in N$ **do**

            forall *parameters/variables* $\in d$ **do**

                Tag the best moth position by flames.

                Update Flame$_{number}$, x and ℓ parameters.

                Calculate moth D – Eq. 22

                Update the matrix M – Eqs. 17 and 20

                if *moths/ants* $\geq$ *Flame$_{number}$* **then**

                    Select an antlion using roulette wheel.

                    Update c and d using – Eqs. 18 and 19

                    Create random walk and normalize it. – Eq. 6

                    Update elite antlion and compare with moth fitness –
Eq. 12

                    Update best antlion/moth position.

                end

            end

            Update elite if an antlion becomes fitter than the elite

        end

        Calculate all fitness values

        Update flames

    until $x \leq$ *Iterations$_{MAX}$*

    Return **Best flame/antlion** position

end

Algorithm 1: MoFAL Framework

where n denotes the number of moths. Another important characteristics of the algorithm are the moth-flames and is given by the below equation:

$$\mathbf{F} = \begin{bmatrix} F_{1,1} & F_{1,2} & F_{1,3} & \dots & F_{1,d} \\ F_{2,1} & F_{2,2} & F_{2,3} & \dots & F_{2,d} \\ \dots & \dots & \dots & \dots & \\ F_{n,1} & F_{n,2} & F_{n,3} & \dots & F_{n,d} \end{bmatrix} \tag{4}$$

For the flames, the corresponding fitness values are stored in the below matrix:

$$\mathbf{OF} = \begin{bmatrix} OF_1 \\ OF_2 \\ \vdots \\ \vdots \\ OF_n \end{bmatrix} \tag{5}$$

Similarly for ALO, the algorithm mimics the interaction between the antlions and ants in the trap. In its mathematical model, ants are required to move over the search space, and antlions are allowed to hunt them and become fitter using traps. Since ants move stochastically in nature when searching for food, a random walk is chosen for modelling ants movement and it is given by:

$$X(t) = [0, \Omega(2\varphi(t_1), \Omega(2\varphi(t_2) - 1), \cdots, \Omega(2\varphi(t_n) - 1)] \tag{6}$$

where Ω calculates the cumulative sum, n is the maximum number of iterations, t shows the step of random walk (iteration , and $\varphi(t)$ is a stochastic function defined by:

$$\varphi(t) = \begin{cases} 1, & \text{if rand} > 0.5 \\ 0, & \text{otherwise} \end{cases} \tag{7}$$

The position of ants are stored and used during optimization in the matrix equation:

$$\mathbf{M_{ant}} = \begin{bmatrix} A_{1,1} & A_{1,2} & A_{1,3} & \dots & A_{1,d} \\ A_{2,1} & A_{2,2} & A_{2,3} & \dots & A_{2,d} \\ \dots & \dots & \dots & \dots & \\ A_{n,1} & A_{n,2} & A_{n,3} & \dots & A_{n,d} \end{bmatrix} \tag{8}$$

The corresponding fitness function are stored in following equation in order to evaluate the fitness of each ant.

$$\mathbf{M_{OA}} = \begin{bmatrix} f([A_{1,1} \ A_{1,2} \ A_{1,3} \ \dots \ A_{1,d}]) \\ f([A_{2,1} \ A_{2,2} \ A_{2,3} \ \dots \ A_{2,d}]) \\ \dots \dots \dots \\ f([A_{n,1} \ A_{n,2} \ A_{n,3} \ \dots \ A_{n,d}]) \end{bmatrix} \tag{9}$$

Furthermore, it is assumed that antlions are hiding at some places in the search space and their positions and fitness solutions are saved in the matrix form:

$$\mathbf{M}_{\text{antlion}} = \begin{bmatrix} Al_{1,1} & Al_{1,2} & Al_{1,3} & \dots & Al_{1,d} \\ Al_{2,1} & Al_{2,2} & Al_{2,3} & \dots & Al_{2,d} \\ \dots\dots\dots\dots\dots \\ Al_{n,1} & Al_{n,2} & Al_{n,3} & \dots & Al_{n,d} \end{bmatrix} \tag{10}$$

where $M_{antlion}$ is the matrix for saving the position of each antlion, $Al_{i,j}$ shows the jth dimension's value of ith antlion, n is the number of antlions, and d is the number of variables (dimension). M_{OAL} is the matrix for saving the fitness of each antlion, $Al_{i,j}$ shows the jth dimension's value of ith antlion, n is the number of antlions, and f is the fitness function.

$$\mathbf{M}_{\text{OAL}} = \begin{bmatrix} f([Al_{1,1} & Al_{1,2} & Al_{1,3} & \dots & Al_{1,d}]) \\ f([Al_{2,1} & Al_{2,2} & Al_{2,3} & \dots & Al_{2,d}]) \\ \dots\dots\dots\dots\dots \\ f([Al_{n,1} & Al_{n,2} & Al_{n,3} & \dots & Al_{n,d}]) \end{bmatrix} \tag{11}$$

In MoFAL, the elitism characteristics of ALO algorithm is extracted and harmonized into MFO algorithm. Elitism is an important characteristic of evolutionary algorithms that allows them to maintain the best fitness(s) obtained at any stage of optimization process. In ALO, the best antlion obtained so far in each iteration is saved and considered as an elite. Since elite is the fittest antlion, it can influence the movements of all ants during iteration. Therefore, it is assumed that every ants random walks around a selected antlion by the roulette wheel and the elite simultaneously as given by the equation:

$$Ant_i^t = \frac{R_A^t + R_E^t}{2} \tag{12}$$

where R_A^t is the random walk around the antlion selected by the roulette wheel at t^{th} iteration, R_E^t is the random walk around the elite at t^{th} iteration, and Ant_i^t indicates the position of ith ant at t^{th} iteration. The ALO algorithm can be deduced to a three-tuple function that search for global minimum for optimization as follows:

$$ALO(A, B, C) \tag{13}$$

where A is a function that generates the random initial solutions, B manipulates the initial population provided by the function A, and C returns true when the end criterion is satisfied. The functions A, B, and C are defined as follows:

$$\theta \xrightarrow{A} \{M_{Ant}, M_{OA}, M_{Antlion}, M_{OAL}\} \tag{14}$$

$$\{M_{Ant}, M_{Antlion}\} \xrightarrow{B} \{M_{Ant}, M_{Antlion}\} \tag{15}$$

$$\{M_{Ant}, M_{Antlion}\} \xrightarrow{C} \{true, false\} \tag{16}$$

where M_{Ant} is the matrix of the position of ants, $M_{Antlion}$ includes the position of antlions, M_{OA} contains the corresponding fitness of ants, and M_{OAL} has the fitness of antlions. It is interesting to note that, MFO is also three-tuple function that approximates the global optimal of optimization problems. This uncanny similarity between ALO and MFO serves as one of the basis of our hybridization in (Siddavaatam and Sedaghat 2021). Furthermore, the position of each moth is updated with respect to a flame using the following equation:

$$M_i = S(M_i, F_j) \tag{17}$$

where M_i indicate the ith moth, F_j indicates the jth flame, and S is the spiral function. For MoFAL, in addition to navigation method of moths, a random walk of antlions is created and normalize it using following equations:

$$c^t = \frac{c^t}{I} \tag{18}$$

$$d^t = \frac{d^t}{I} \tag{19}$$

where I is a ratio, c^t is the minimum of all variables at t^{th} iteration, and d^t indicates the vector including the maximum of all variables at t^{th} iteration. Elite antlion position is calculated and compared with that of moth position as given by the logarithmic spiral equation below:

$$\overrightarrow{X(t+1)} = \overrightarrow{D'} \cdot e^{bl} \cdot \cos(2\pi l) + \overrightarrow{X^*(t)} \tag{20}$$

where $\overrightarrow{D'} = [D_i]$ indicates the distance of the ith moth for the jth flame, b is a constant for defining the shape of the logarithmic spiral, and ℓ is a random number in $[y, 1]$ where adaptive convergence y linearly decreases from -1 to -2 over the course of several iterations. Flame number (Flame$_{number}$) is calculated by following equation:

$$\text{Flame}_{\text{number}} = round\left((W_{\text{n}} - \text{Iteration}) * \frac{W_{\text{n}} - 1}{\text{Iterations}_{\text{maximum}}}\right) \tag{21}$$

where W_n is the number of search agents. D is deduced as follows:

$$D_i = |F_j - M_i| \tag{22}$$

If Elite antlion fitness is greater than that of moth-flame fitness; position vectors are updated using Eq. 12 using both moth flame position and elite ant-lion position.

3   Model Architecture

In Sect. 3.1, an efficient CNN in the form of MobileNetV3 was discussed and is utilized as models to perform image recognition tasks where they can act as a core component in the feature extraction phase. The MobileNetV3 is used to pretrain a model using the ImageNet dataset in order to avoid retraining the model from the initial state. This pretrained model was adopted to COVID-19 recognition task via transfer learning and fine-tuning as part of the experiments. A standard procedure to fine-tune the model is applied and extract the relevant image embeddings in dual stages. In the first stage, the model undergoes changes in the top two layers of the model utilized for image classification with a 1×1 point-wise convolution for extracting features from images. The convolution feature can be regarded as a multilayer perceptron (MLP) which assists in reduction of dimensionality or performs classification of images by integration of different non-linearity operations. Furthermore, other 1×1 point-wise convolutions have been added on the top of the model for fine-tuning the model's weights on different datasets related to the classification task. In the second stage following the fine-tuning process, the output of the 1×1 point-wise convolution is flattened and deployed for feature extraction to generate image embeddings with the size of 128 for each image in the dataset. Finally, the feature selection phase can utilize the extracted image embeddings that are routed to it.

3.1   MobileNetV3

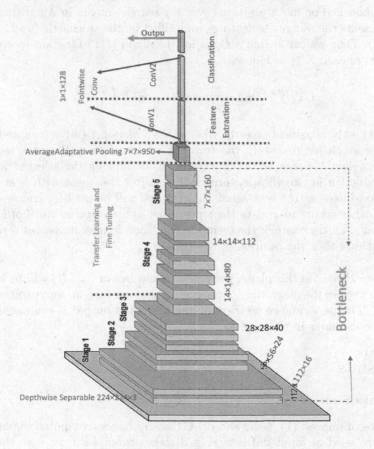

Fig. 2. MobileNetV3 archictecture

Figure 2 illustrates the variation of the MobileNetV3 employed for COVID-19 images feature extraction. In this variation, the feature extraction phase is performed after fine-tuning the model for 100 epochs during ten randomly initialized runs, thereby resulting in the highest classification accuracy on each dataset. A stochastic gradient descent method named RMSprop was deployed to fine-tune the model with a batch size of 32 at a learning rate set to 1×10^{-4}. The overfitting problem was mitigated using data augmentation in the data preprocessing resulting in an improvement of model's generalization. Techniques like random crop, random horizontal flip, color jittery, and random vertical flip were some of the data augmentation transformations utilized by juxtaposing with original image resizing to shape 224×224.

Learning Phase. In this stage 71% of the dataset was utilized as a training set to learn the model thus leading to effective feature selection process. Consequently, the primary hyrbid approach is to set the initial values for the population $X(t)$ determined by either Eq. (6) modelled on the stochastic movement of ants or Eq. (22) modelled on moth positions over the search domain in Algorithm 1 and W_n represents the number of features quantified by the stochastic function $\varphi(t)$ in Eq. (7). Thus we can define a classifier-threshold (CT) function to compute the fitness of each $X(t)$ as follows:

$$\mathrm{CT}(t) = \kappa * \theta(t) + (1 - \kappa) * \left(\frac{\varphi(t)}{W_n} \right) \tag{23}$$

where $\theta(t)$ is the classification error when a KNN classifier (Silverman and Jones 1989) was employed to reduce the training set based on $\varphi(t)$ and κ serves as a counterweight between the bi-objectives of minimizing the selected features versus reduction in classification errors. Thereafter the agent with best fitness trait is considered as the best agent $X_{best}(t)$ and will be used for course corrections on other agents to update their positions as discussed in steps of MoFAL Algorithm 1. Furthermore, if the terminal condition of max iterations is reached in Algorithm 1 then the position of is returned.

Evaluation Phase. In this phase, the relevant features in $X_{best}(t)$ will be utilized to further reduce the magnitude of testing set and use it as an input to the KNN classifier. The performance metrics for the predicted output is evaluated using various benchmarks in the next section.

4   Results

4.1   Data Sources

Two types of images: (1) X-ray and (2) CT scan images (computed tomography scan) were used as input datasets (Fig. 3 shows examples from each dataset). During the course of our experiments on the model, we utilized three different datasets to train and fine-tune the feature extraction model: (1) COVID-CT

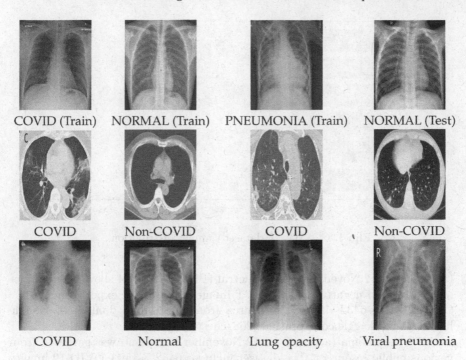

COVID (Train)   NORMAL (Train)   PNEUMONIA (Train)   NORMAL (Test)

COVID   Non-COVID   COVID   Non-COVID

COVID   Normal   Lung opacity   Viral pneumonia

Fig. 3. COVID-XRay-6432, COVID-CT and COVID-19 radiography dataset samples in 1st, 2nd and 3rd rows respectively

dataset (dataset1), (2) COVID-XRay-6432 dataset (dataset2), and (3) COVID-19 radiography dataset (dataset3). Furthermore, the same data split is maintained after extracting image embeddings from each dataset which are fed to a feature selection and classification phase. More details on the datasets are described as follows:

COVID-CT dataset: is sourced from two places, including research papers (for training) and original CT scans donated by hospitals (for testing). For the research papers, a collection (Yang et al. 2020) of 760 preprints from two databases was used including medRxiv https://www.medrxiv.org/ (accessed on 12 November 2021) and bioRxiv https://www.biorxiv.org/ (accessed on 12 October 2021). The preprints were collected from papers posted from 19 January to 25 May 2021. In total, 349 CT images labeled as positive were collected from 216 patient cases for COVID-19. In addition, the authors collected 397 negative CT images (Non-Covid19) to build their dataset for a binary classification task from sources including MedPix https://medpix.nlm.nih.gov/home (accessed on 12 November 2021) database, the LUNA7 https://luna16.grand-challenge.org/ (accessed on 12 November 2021) dataset, the Radiopaedia https://radiopaedia.org/articles/covid-19-3 (accessed on 12 November 2021) website, and PubMed https://www.ncbi.nlm.nih.gov/pmc/

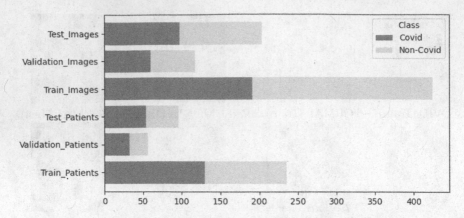

Fig. 4. COVID-CT dataset samples distribution.

(accessed on 12 November 2021) Central (PMC). Figure 4 shows the number of positive and negative Covid-19 CT images used in our experiments.

- COVID-XRay-6432 dataset: is source from the public domain of Kaggle https://www.kaggle.com/prashant268/chest-xray-covid19-pneumonia (accessed on 12 November 2021) and was gathered from various public resources. The dataset includes 6432 X-ray COVID-19 images distributed on three classes which are COVID-19, PNEUMONIA, and NORMAL (Non-COVID). The training set comprises 80% of the dataset, and the test set comprises 20% of the dataset. In our experiments, 15% of the training sample is used in the validation set and fine-tuning. The number of samples in each class is shown in Fig. 5.
- COVID-19 radiography dataset: is collected by a team of researchers from different countries and universities collaborating with medical professionals. It is an opensource available online and frequently updated on Kaggle https://www.kaggle.com/tawsifurrahman/covid19-radiography (accessed on 12 November 2021). The dataset consists of 21,165 chest X-ray (CXR) COVID-19 images distributed on four categories which are COVID19, lung opacity, viral pneumonia, and NORMAL (Non-COVID). In our experiment, we randomly split the data into 70%, 10%, and 20% for training, validation, and testing sets, respectively. The number of samples in each class after splitting the data is listed in Fig. 6.

4.2 Metrics

Accuracy (Acc) estimation for the model was done using a few statistical parameters as the mean of the worst values (Max), mean of best values, standard deviation, and computational time elapsed during the selection of features including the classification phase. The measures are outlined as follows:

$$\mathrm{Acc} = \frac{\mathrm{TP} + \mathrm{TN}}{\mathrm{TP} + \mathrm{TN} + \mathrm{FP} + \mathrm{FN}} \tag{24}$$

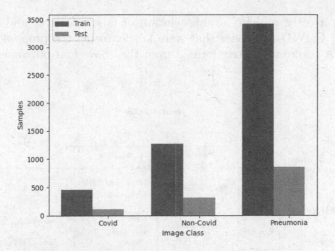

Fig. 5. COVID-XRay-6432 dataset samples distribution.

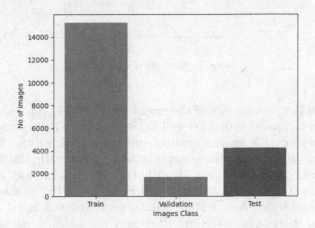

Fig. 6. COVID-19 radiography dataset samples distribution.

$$\text{Sensitivity} = \frac{TP}{TP + FN} \tag{25}$$

$$\text{Specificity} = \frac{TN}{TN + FP} \tag{26}$$

$$F_{score} = 2 \times \frac{\text{Sensitivity} \times \text{Specificity}}{\text{Sensitivity} + \text{Specificity}} \tag{27}$$

where TP – abbreviation of true positives and represents the positive COVID-19 images labeled using the proposed classifier correctly, TN – stands for the true negative samples and represents the negative COVID-19 images that were labeled using the proposed classifier correctly, FP – abbreviation of false positives and represents the positive COVID-19 images labeled using the proposed classifier incorrectly, and FN – abbreviation of false negatives and represents the negative COVID-19 images that were labeled using the proposed classifier incorrectly. A confusion matrix formed from the above four outcomes is shown in Fig. 7.

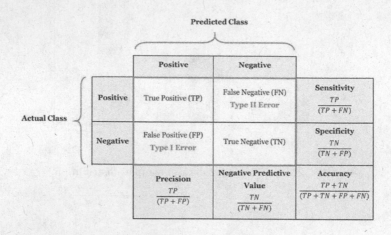

Fig. 7. Confusion matrix

To validate the performance of the model with MoFAL as a feature selector, its results were compared with other well-known Feature selectors based on Meta-heuristic techniques. For example, whale optimization algorithm (WOA) (Mirjalili and Lewis 2016), moth-flame optimization (MFO) (Mirjalili 2015), firefly algorithm (FFA) (Yang 2009), bat algorithm (BAT) (Huang et al. 2017c), Grey Wolf algorithm (GWO) (Siddavaatam and Sedaghat 2019), simulated annealing. The parameters of these feature selectors are assigned based on the original implementation of each method. However, the common parameters, such as the number of iterations and population size, are set to 50 and 15, respectively. In addition, each feature selector conducted 25 runs for a fair comparison between them. All DL training and feature extraction phases were conducted on a GPU (Graphics processing unit) from Nvidia, while the feature selection phase was experimented on the Google collaboratory platform. For a proper validation of the framework, other DL models such as DenseNet (Huang et al. 2017), VGG19 (Simonyan and Zisserman 2015), and EfficientNet (Tan and Le 2019) were exploited as backbone feature extraction methods using their standard architecture and parameters described in the Sect. 4.3 using comparative heatmaps in Figs. 12 and 13.

4.3   Analysis

Two datasets are used to evaluate the performance of the model and the findings are summarized in Table 1 and Figs. 7, 11, 9 and 8. Specifically, Table 1 can be

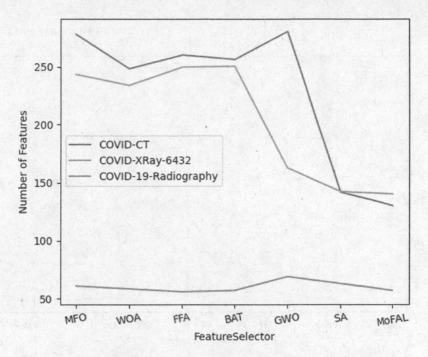

Fig. 8. Number of features selected in every dataset

Table 1. Statistical measures for MoFAL comparison with similar metaheurisitc benchmarks

	Measure	MFO	WOA	FFA	BAT	GWO	SA	MOFAL
(1) ref 4.1	Accuracy	0.769	0.761	0.769	0.771	0.773	0.766	0.883
	Recall	0.761	0.765	0.767	0.771	0.778	0.752	0.881
	Precision	0.771	0.764	0.771	0.775	0.776	0.770	0.885
	F1-Score	0.767	0.76	0.767	0.769	0.772	0.764	0.882
(2)	Accuracy	0.857	0.925	0.858	0.863	0.863	0.858	0.874
	Recall	0.856	0.82	0.812	0.744	0.658	0.827	0.894
	Precision	0.828	0.867	0.873	0.802	0.859	0.867	0.914
	F1-Score	0.842	0.843	0.841	0.867	0.859	0.892	0.910

used to evaluate the performance of a model and determine the classification accuracy among the first two tested datasets. The MoFAL optimizer model was evaluated for classification tasks on both the datasets and it outperforms many state-of-the-art baselines. Furthermore, the accuracy of MoFAL optimizer is comparatively better for the datasets mentioned in Table 1. In addition, MoFAL has a higher Recall, Precision, and F1 score overall than other feature selection

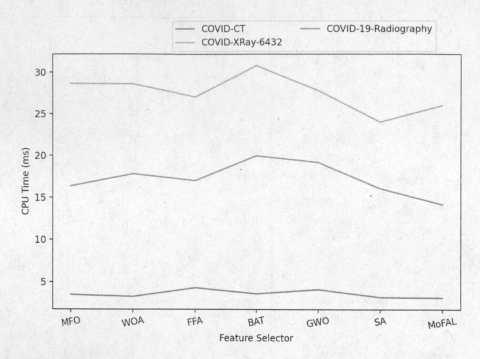

Fig. 9. CPU time consumed for each type of feature selector w.r.t every dataset

methods. As in the case of dataset1, the efficiency of MoFAL using dataset2 is the best in terms of classification accuracy, followed by BAT and GWO. We notice that the recall value of GWO and MFO is comparatively better than all other methods (i.e., WOA, FFA, BAT, and GWO) except for MoFAL algorithm, as well as the precision value which ranks MoFAL as the best among comparators. Similarly, with regard to F1-Scores, the MoFAL significantly performs better when compared with other methods described in Fig. 10

The average of each algorithm for the traits of accuracy, recall, precision, and F1-score is depicted in Fig. 10 which shows that MoFAL approach is better approach than other comparable algorithms. Furthermore, computational time taken in CPU milliseconds is described in Fig. 9 which clearly shows the advantage of our framework over other comparable methods for every type of dataset. Also, the efficiency of MoFAL to reduce the number of features is observed from the number of selections made on features. MoFAL has the smallest number of features, 130 and 140 at Dataset1 and Dataset2, respectively. Figure 11 which shows the average over the two datasets in terms of CPU time in milliseconds (ms) and number of Feature selections, MoFAL performs better in these aspects as well.

The performance of the model that combines the MobileNetV3 and MoFAL is significant improvement when compared with the other three CNN types networks. These networks include VGG19 (Simonyan and Zisserman 2015) (Visual

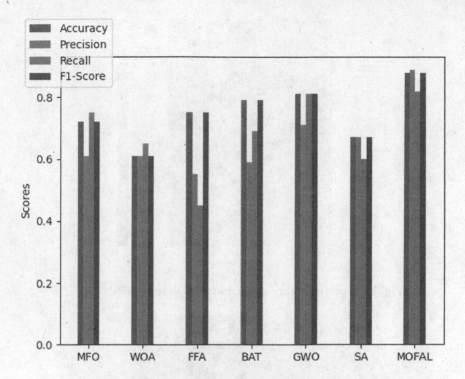

Fig. 10. Interpretation of performance measures

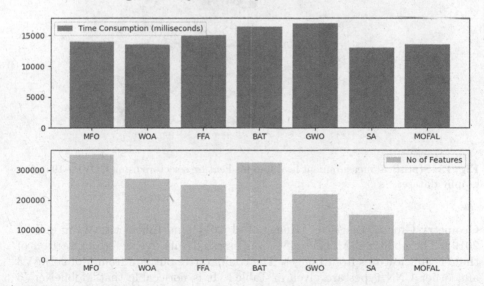

Fig. 11. Average CPU time consumption and number of features selected for respective algorithms

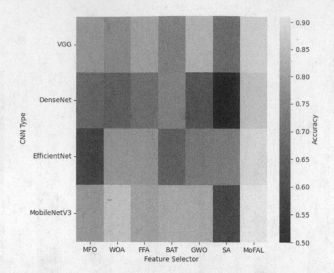

Fig. 12. Accuracy heatmap for feature selector vs CNN type.

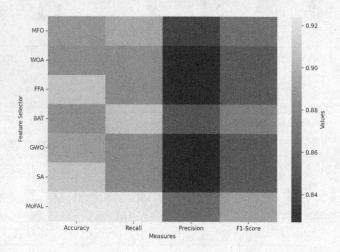

Fig. 13. Statistical measurement heatmap for Feature selectors using COVID-19 radiography dataset.

Geometry Group), DenseNet (Huang et al. 2017) and EfficientNet (Tan and Le 2019). The ability of MobileNetV3 was assessed with respect to extraction of the relevant features from Images. The comparative outputs from MobileNetV3 and other CNN types are given in Table 1. It is noticeable that MobileNetV3 can provide better performance than other CNNs closely followed by DenseNet that has a high ability to extract relevant features better than the other two networks (i.e., VGG and EfficientNet). A similar observation is shown in Fig. 12 which estimates that the average of the accuracy in all the tested Feature Selection algorithms using the feature extracted from each CNN type is better. The performance of MoFAL based on MobileNetV3 in terms of accuracy among the two tested datasets is described in Fig. 8. From these averages, it can be deduced that the developed model provides better results than other comparable CNNs. In addition, the ability of MoFAL to increase the accuracy classification is better than other feature selectors when using different CNN types.

A load test is performed using the third COVID-19 radiography dataset for assessing the impact of handling large number of images. Figure 13 shows the statistical measured outcomes in terms of performance for each feature selector that depends on the features extracted using MobileNetV3 for COVID-19 radiography dataset. We can further conclude that our model has a comparatively better capacity to enhance the classification accuracy and reduce the number of features required to increase the classification accuracy.

Conclusion

This article proposes a new model framework using MobileNetV3 DL model and MoFAL optimizer to detect the COVID-19 cases from X-ray and CT images using three datasets with a significant amount of samples. Extensive comparisons were made with three other state of the art DL networks namely VGG19, DenseNet, and EfficientNet. The extracted image embeddings from each DL network were fed to the feature selector for feature space reduction and classification performance improvement. Three large and recent datasets were used with different characteristics since they represent X-ray and CT COVID-19 images collected from multiple and varied sources. The comparison results illustrated the high performance of the developed method based on our developed model over the other competitive approaches. In addition, our model can be easily retrofitted to other applications such as geospatial, astrophysics, and other image classification tasks. The CPU times described for various scenarios in Figs. 9 and 11 can be further improvised using python based multithreaded approach. As part of future work, we plan to harness multi-threaded programming to further improve this particular aspect of the model described in this article.

Acknowledgement. I would like to acknowledge my late mother Ms. Meena Ramaiah, who inspired me to research on COVID-19 in this article.

References

Ahmed, Z., Mohamed, K., Zeeshan, S., Dong, X.: Artificial intelligence with multifunctional machine learning platform development for better healthcare and precision medicine. Database 2020, January 2020

Bernheim, A., Mei, X., Huang, M., Yang, Y., Fayad, Z.A., Zhang, N.: Chest CT findings in coronavirus disease-19 (COVID-19): relationship to duration of infection. Radiology **295**(3), 200463 (2020)

Boberg-Fazlic, N., Lampe, M., Pedersen, M., Sharp, P.: Pandemics and protectionism: evidence from the "Spanish" flu. Humanit. Soc. Sci. Commun. **8**, 1–9 (2021)

Canayaz, M.: MH-COVIDNet: diagnosis of COVID-19 using deep neural networks and meta-heuristic-based feature selection on x-ray images. Biomed. Sig. Process. Control **64**, 102257 (2021)

Chaddad, A., Hassan, L., Desrosiers, C.: Deep CNN models for predicting COVID-19 in CT and x-ray images. J. Med. Imaging (Bellingham) **8**(S1), 014502 (2021)

Ciga, O., Xu, T., Nofech-Mozes, S., Noy, S., Lu, F.I., Martel, A.L.: Overcoming the limitations of patch-based learning to detect cancer in whole slide images. Sci. Rep. **11**(1), 8894 (2021)

Dai, W.C., et al.: CT imaging and differential diagnosis of COVID-19. Can. Assoc. Radiol. J. **71**, 195–200 (2020)

Demirci, N.Y., et al.: Relationship between chest computed tomography findings and clinical conditions of coronavirus disease (COVID-19): a multicentre experience. Int. J. Clin. Pract. **75**(9), e14459 (2021)

Elaziz, M.A., et al.: An improved marine predators algorithm with fuzzy entropy for multi-level thresholding: real world example of COVID-19 CT image segmentation. IEEE Access **8**, 125306–125330 (2020)

Harjoseputro, Y., Yuda, I.P., Danukusumo, K.P.: MobileNets: efficient convolutional neural network for identification of protected birds. Int. J. Adv. Sci. Eng. Inf. Technol. **10**(6), 2290 (2020)

Howard, A., Sandler, M., Chen, B., Wang, W., Chen, L.C.: Searching for MobileNetV3. In: 2019 IEEE/CVF International Conference on Computer Vision (ICCV). IEEE, October 2019

Huang, G., Liu, Z., Maaten, L.V.D., Weinberger, K.Q.: Densely connected convolutional networks. In: 2017 IEEE Conference on Computer Vision and Pattern Recognition (CVPR), pp. 2261–2269. IEEE, July 2017

Huang, X., Zeng, X., Han, R.: Dynamic inertia weight binary bat algorithm with neighborhood search. Comput. Intell. Neurosci. **2017**, 1–15 (2017). https://doi.org/10.1155/2F2017/2F3235720

Ignatov, A., et al.: Real-time video super-resolution on smartphones with deep learning, mobile AI 2021 challenge: report. In: 2021 IEEE/CVF Conference on Computer Vision and Pattern Recognition Workshops (CVPRW). IEEE, June 2021

Ji, J., Krishna, R., Fei-Fei, L., Niebles, J.C.: Action genome: actions as compositions of spatio-temporal scene graphs. In: 2020 IEEE/CVF Conference on Computer Vision and Pattern Recognition (CVPR). IEEE, June 2020

Kesim, E., Dokur, Z., Olmez, T.: X-ray chest image classification by a small-sized convolutional neural network. In: 2019 Scientific Meeting on Electrical-Electronics Biomedical Engineering and Computer Science (EBBT), pp. 1–5 (2019)

Khuzani, A.Z., Heidari, M., Shariati, S.A.: COVID-classifier: an automated machine learning model to assist in the diagnosis of COVID-19 infection in chest x-ray images. Sci. Rep. **11**(1), 9887 (2021)

Kundu, R., Basak, H., Singh, P.K., Ahmadian, A., Ferrara, M., Sarkar, R.: Fuzzy rank-based fusion of CNN models using gompertz function for screening COVID-19 CT-scans. Sci. Rep. **11**(1), 14133 (2021)

Larici, A.R., Cicchetti, G., Marano, R., Bonomo, L., Storto, M.L.: COVID-19 pneumonia: current evidence of chest imaging features, evolution and prognosis. Chin. J. Acad. Radiol. **4**, 229–240 (2021)

Liu, J., Inkawhich, N., Nina, O., Timofte, R.: NTIRE 2021 multi-modal aerial view object classification challenge. In: Proceedings of the IEEE/CVF Conference on Computer Vision and Pattern Recognition (CVPR) Workshops, pp. 588–595, June 2021

Maior, C.B.S., Santana, J.M.M., Lins, I.D., Moura, M.J.C.: Convolutional neural network model based on radiological images to support COVID-19 diagnosis: evaluating database biases. PLoS ONE **16**(3), e0247839 (2021)

Mirjalili, S.: Moth-flame optimization algorithm: a novel nature-inspired heuristic paradigm. Knowl. Based Syst. **89**, 228–249 (2015). https://doi.org/10.1016/2Fj.knosys.2015.07.006

Mirjalili, S., Lewis, A.: The whale optimization algorithm. Adv. Eng. Softw. **95**, 51–67 (2016). https://doi.org/10.1016/j.advengsoft.2016.01.008

Narin, A.: Accurate detection of COVID-19 using deep features based on x-ray images and feature selection methods. Comput. Biol. Med. **137**, 104771 (2021)

Oh, Y., Park, S., Ye, J.C.: Deep learning COVID-19 features on CXR using limited training data sets. IEEE Trans. Med. Imaging **39**(8), 2688–2700 (2020)

Onder, O., Yarasir, Y., Azizova, A., Durhan, G., Onur, M.R., Ariyurek, O.M.: Errors, discrepancies and underlying bias in radiology with case examples: a pictorial review. Insights Imaging **12**(1), 51 (2021)

Ramachandran, P., Zoph, B., Le, Q.V.: Searching for activation functions. CoRR abs/1710.05941 (2017). https://arxiv.org/abs/1710.05941

Roy, S., et al.: Deep learning for classification and localization of COVID-19 markers in point-of-care lung ultrasound. IEEE Trans. Med. Imaging **39**(8), 2676–2687 (2020)

Sahlol, A.T., Yousri, D., Ewees, A.A., Al-qaness, M.A.A., Damasevicius, R., Elaziz, M.A.: COVID-19 image classification using deep features and fractional-order marine predators algorithm. Sci. Rep. **10**(1), 15364 (2020)

Siddavaatam, P., Sedaghat, R.: Grey wolf optimizer driven design space exploration: a novel framework for multi-objective trade-off in architectural synthesis. Swarm Evol. Comput. **49**, 44–61 (2019)

Siddavaatam, P., Sedaghat, R.: A new bio-heuristic hybrid optimization for constrained continuous problems. In: Gavrilova, M.L., Tan, C.J.K. (eds.) Transactions on Computational Science XXXVIII. LNCS, vol. 12620, pp. 76–97. Springer, Heidelberg (2021). https://doi.org/10.1007/978-3-662-63170-6_5

Silverman, B.W., Jones, M.C.: E. Fix and J.L. Hodges (1951): an important contribution to nonparametric discriminant analysis and density estimation. Commentary on Fix and Hodges (1951). Int. Stat. Rev./Revue Internationale de Statistique **57**(3), 233 (1989)

Simonyan, K., Zisserman, A.: Very deep convolutional networks for large-scale image recognition (2015)

Sun, L., Shao, W., Wang, M., Zhang, D., Liu, M.: High-order feature learning for multi-atlas based label fusion: application to brain segmentation with MRI. IEEE Trans. Image Process. **29**, 2702–2713 (2020)

Tan, M., et al.: MnasNet: platform-aware neural architecture search for mobile. In: 2019 IEEE/CVF Conference on Computer Vision and Pattern Recognition (CVPR). IEEE, June 2019

Tan, M., Le, Q.: EfficientNet: rethinking model scaling for convolutional neural networks. In: Chaudhuri, K., Salakhutdinov, R. (eds.) Proceedings of the 36th International Conference on Machine Learning, 09–15 June 2019, vol. 97, pp. 6105–6114. Proceedings of Machine Learning Research. PMLR (2019)

Tran, D., Wang, H., Feiszli, M., Torresani, L.: Video classification with channel-separated convolutional networks. In: 2019 IEEE/CVF International Conference on Computer Vision (ICCV). IEEE, October 2019

Wang, J., et al.: Prior-attention residual learning for more discriminative COVID-19 screening in CT images. IEEE Trans. Med. Imaging **39**(8), 2572–2583 (2020)

Wang, S., et al.: A fully automatic deep learning system for COVID-19 diagnostic and prognostic analysis. Eur. Respir. J. **56**(2), 2000775 (2020)

Wang, X., et al.: A weakly-supervised framework for COVID-19 classification and lesion localization from chest CT. IEEE Trans. Med. Imaging **39**(8), 2615–2625 (2020)

Yang, X.S.: Firefly algorithm, lévy flights and global optimization. In: Bramer, M., Ellis, R., Petridis, M. (eds.) Research and Development in Intelligent Systems XXVI. Springer, London (2010). https://doi.org/10.1007/978-1-84882-983-1_15

Yang, X., He, X., Zhao, J., Zhang, Y., Zhang, S., Xie, P.: COVID-CT-dataset: a CT scan dataset about COVID-19 (2020)

Yousri, D., Elaziz, M.A., Abualigah, L., Oliva, D., Al-qaness, M.A., Ewees, A.A.: COVID-19 x-ray images classification based on enhanced fractional-order cuckoo search optimizer using heavy-tailed distributions. Appl. Soft Comput. **101**, 107052 (2021)

Zhang, K., et al.: Clinically applicable AI system for accurate diagnosis, quantitative measurements, and prognosis of COVID-19 pneumonia using computed tomography. Cell **181**(6), 1423.e11–1433.e11 (2020)

Zhang, X., Zhou, X., Lin, M., Sun, J.: ShuffleNet: an extremely efficient convolutional neural network for mobile devices. In: 2018 IEEE/CVF Conference on Computer Vision and Pattern Recognition. IEEE, June 2018

Zhu, Z., Lian, X., Su, X., Wu, W., Marraro, G., Zeng, Y.: From SARS and MERS to COVID-19: a brief summary and comparison of severe acute respiratory infections caused by three highly pathogenic human coronaviruses. Respir. Res. **21**, 224 (2020)

Zoph, B., Vasudevan, V., Shlens, J., Le, Q.V.: Learning transferable architectures for scalable image recognition. In: 2018 IEEE/CVF Conference on Computer Vision and Pattern Recognition. IEEE, June 2018

An Unsupervised DNN Embedding System for Image Clustering

Abu Quwsar Ohi[(✉)] [ID]

Bangladesh University of Business and Technology, Dhaka, Bangladesh
quwsarohi@bubt.edu.bd

Abstract. Recently, AutoEmbedder has been introduced that implements DNN classifiers as an embedding system. Although the AutoEmbedder generates clusterable embeddings, it is trained in a supervised approach. Therefore, in this paper, we introduce an unsupervised AutoEmbedder (UAutoEmbedder) that is trained in an unsupervised fashion. Through rigorous research effort, we discovered that AutoEmbedder is an ideal architecture for data augmentation dependent unsupervised learning. Hence, we propose a random sampling scheme and perform image data augmentation to train the AutoEmbedder system in an unsupervised manner. Furthermore, we reform the training procedure of the AutoEmbedder that properly utilizes the unsupervised learning strategy. We evaluate the model by applying different datasets and conduct an in-depth study of the generated results. From the comprehensive evaluations, we perceive that UAutoEmbedder has enormous opportunities in unsupervised learning and image data augmentation.

Keywords: Deep neural network · Unsupervised learning · Embedding · Clustering · Dimensionality reduction · Pairwise constraint · Image augmentation

1 Introduction

Clustering is the process of separating data based on some common properties. In the concept of Machine Learning, clustering is mostly considered as an unsupervised learning strategy that explores relationships among numerous unknown data. In computer vision, unsupervised learning is continuously being improved using the fusion of clustering and Deep Neural Network (DNN) architectures. Popular DNN strategies such as Convolutional Deep Neural Networks (CDNN), Generative Adversarial Network (GAN), and AutoEncoder (AE) are widely studied and fused with clustering methods to generate unsupervised image recognition methods. The fusion of DNN and unsupervised strategies is often regarded as Deep Clustering (DC).

In the present era of the internet-depended social media, unlabelled data can be gathered easily through crawling. To develop an intelligent system, these

© Springer-Verlag GmbH Germany, part of Springer Nature 2022
M. L. Gavrilova and C. J. K. Tan (Eds.): Trans. on Comput. Sci. XXXIX, LNCS 13460, pp. 109–126, 2022.
https://doi.org/10.1007/978-3-662-66491-9_6

crawled data must be again labeled manually to feed the autonomous system. However, data labeling is one of the main challenges of implementing an autonomous system, which requires manual labor and therefore, it is also cost-consuming. As unsupervised learning strategies do not require data labeling, they can reduce the production cost of classification and clustering models.

There exist popular clustering algorithms to find linear relationships among data, such as k-means [17], Density-based spatial clustering of applications with noise (DBSCAN) [8], hierarchical clustering [18], etc. However, these clustering schemes have common limitations: 1) These methods operate better on low dimensional data. 2) These methods operate better on linear features, yet, most data does not contain linear features. The limitations can be solved if it is possible to generate a function that can convert high dimensional data into low dimensional data while converting the higher dimensional relation into a linear relationship. Currently, DNN architectures are superior in recognizing high dimensional features. Therefore, DNN architectures are being used to reduce the dimension of data. Almost all of the current DC strategies follow the scheme of utilizing DNN to reduce the dimension of data.

Currently, numerous strategies are applied in image clustering. Image clustering is now generated in a supervised, semi-supervised, and unsupervised approach. However, some unsupervised methods require pre-training on similar unlabeled data [34]. In some cases, unsupervised methods are pre-trained on a labelled dataset [14]. Although numerous sophisticated strategies are available in the image DC domain, however, the unsupervised approach has some limitations on the feature selection process. Often, unsupervised procedures can not discover on what basis two images are considered to be different. Therefore, providing no target to the DC method confuses CDNN architectures to select the separable features. Due to the lack or mismatch of this common intuition, most DC methods fail to cluster image datasets correctly.

Image augmentation is a widely used strategy in image classification tasks and is proven to improve the robustness of image classifiers [31]. Through properly augmenting an image, it is possible to reproduce a nearly different image while keeping the distinguishable features. Although image augmentations are widely used in supervised classification tasks, the usage of augmentation in DC strategies are hardly observed [12].

The paper proposes UAutoEmbedder, an improved version of the AutoEmbedder [25] architecture. Generally, AutoEmbedder is trained based on pairwise constraints, and due to the less requirement of training data, the AutoEmbedder is treated as a semi-supervised method. Pairwise constraints are generally considered as a binary condition that defines if two images belong to the same cluster or not. Instead of depending on pre-defined labels of pairwise constraints, we apply simple randomization and augmentation to define pairwise constraints. To train the UAutoEmbedder in an unsupervised approach, we specify that a clean image and its augmented image variant belong to the same cluster. In contrast, we assume a randomly selected pair from the dataset will always belong to different clusters. Using the straightforward hypotheses, we validate the strategy exhibits significant performance improvement on some renowned datasets. U-vector [22] closely resembles the present work applied speech information.

The contributions of the paper are stated below:

- We exploit an unsupervised learning scheme of the AutoEmbedder architecture. We use simple augmentation and random selection policy to define the pairwise relation of data.
- We improve the primary training method and architectural policy of the AutoEmbedder.
- We execute experiments on some well-known datasets and validate that the proposed method performs better in some of the complex datasets.
- We discuss the challenges and opportunities of unsupervised learning using augmentation.

The rest of the paper is constructed as follows. Section 2 outlines the efforts that have been performed in the scope of unsupervised DC. Section 3 acquaints the semi-supervised AutoEmbedder architecture and exhibits the unsupervised training methodology. Section 4 notifies the metrics, datasets and the environmental setup that are used to conduct the evaluation. Section 5 presents the empirical results of the UAutoEmbedder along with the comparison with different unsupervised systems. Section 6 presents a discussion addressing the applications and data dependency of UAutoEmbedder. Finally, Sect. 7 concludes the paper.

2   Related Work

Currently, DC strategies are being widely studied, and numerous methods are being introduced. Generally, AE, GAN, and CDNN are widely observed to exploit DC. Furthermore, a probabilistic manner of AE defined Variational AE (VAE) [20] is also commonly studied in the field of DC.

GAN [11] based architectures generally depend on two CDNN functions, a generator, and a discriminator. The GAN is trained based on a min-max policy. However, to use such design in DC, some modifications are often introduced. ClusterGAN [23] fuses an encoder model with the output of the generator. Through joint training, the encoder learns to cluster the input images. On the contrary, CatGAN [32] and InfoGAN [5] mutate the recognition procedure of the discriminator. GAN based unsupervised and semi-supervised methods still suffer from selecting generator and discriminator architectures [6]. Furthermore, the computational complexity of GAN based methods is comparatively high.

In contrast, AE based architectures combine encoding $f_\phi(\cdot)$ and decoding $g_\theta(\cdot)$ functions to discover proper coding for a given set of data, by minimizing the reconstruction loss of the AE, $L_r = ||x - g_\theta(f_\phi(x))||^2$. AE architectures are formulated in DC in the concept of merging the reconstruction loss of the decoding function with the consideration of a clustering loss. Deep Embedded Clustering (DEC) [34] is one of the most well-known DC methods that comprises AE with a clustering loss. Inspired by the unsupervised strategy of generating clusters using DEC, a semi-supervised strategy of DEC is also introduced [27]. However, some AE architectures do not include clustering loss. DEeP Embedded RegularIzed ClusTering (DEPICT) [10] merges softmax layer cross-entropy loss with AE. Furthermore, it includes a noisy AE to improve the robustness of the architecture. AE is a widely studied and accepted strategy in DC, since compared to GAN strategies, they are computationally less complex. But, due to the limited depth of AE based architecture, often they are lack of extracting high dimensional features.

CDNN based architectures are widely used in DC. CDNN architectures include popular DC strategies that include various cluster loss strategies, such as triplet loss [24], pairwise loss [3], agglomerative clustering [35] etc. In comparison to the AE and GAN based strategies, CDNN architectures do not depend on reconstruction loss. Instead, CDNN architectures directly operate on decreasing the clustering loss.

Data augmentation is not extensively explored in the expanse of the DC method. Information Maximizing Self-Augmented Training (IMSAT) [15] combines a fully connected network with regularized information maximization. IMSAT learns mutual information using probabilistic classifiers. Furthermore, to improve the regularization of the system, it is implemented with self augmented training. The method applies affine transformations for augmenting an image. Deep Embedded Clustering with Data Augmentation (DEC-DA) [12] is an improved version of the DEC architecture that depends on AE. DEC-DA implements an extra encoder with the traditional encoder-decoder approach of the AE. The additional encoder (shared weights with existing encoder) is used to flow the clean image data. In contrast, the default encoder is used to stream an augmented version of the real image data. Only the augmented encoder and decoder are trained via backpropagation. The DEC-DA approach implies that including augmentation can improve the clustering accuracy.

Random Triplet Mining (RTM) [24] is implemented based on a triplet loss [29] architecture. A triplet loss strategy optimizes the clustering criteria based on the embeddings generated from three different inputs. The three input is formally defined as the anchor (α), positive (β), and negative (γ). The positive input belongs to the same cluster as the anchor, whereas the negative input belongs to a different cluster. For a given margin m, the goal of the architecture is to lessen the loss, $L_t = \sum_i^N max(\delta_i^\beta - \delta_i^\gamma + m, 0)$. The RTM directly depends on the augmented images based on the assumption that the clean image and the augmented image must belong to the same cluster, whereas a randomly selected image must belong to a different cluster. However, the method depends on a hyper-parameter d that explains the index of the randomly selected pair of data. The complexity raises selecting the optimal value of d. Because without selecting the optimal value of d, it is often observed that the model does not improve the clustering accuracy. Moreover, the value of d may differ based on the model and the weights of the model.

This paper applies a pairwise loss to generate unsupervised clustering of images. The method implements the basic strategy of AutoEmbedder [25] architecture (explained in Sect. 3.1). Generally, the training strategy of AutoEmbedder is supervised, and the training depends on pairwise constraints. The paper further enhances the supervised training process in a fully unsupervised fashion. The improvement is accomplished based on the general intuition that a pair of clean data and augmented data must belong to the same cluster whereas, a pair of randomly selected images should belong to a different cluster.

3   Method

The proposed strategy of unsupervised training of the UAutoEmbedder involves understanding the basic intuition of the AutoEmbedder architecture. Therefore, in Sect. 3.1, we express the general concept of AutoEmbedder architecture. Consequently, in Sect. 3.2, we represent the unsupervised training strategy of the UAutoEmbedder. In Sect. 3.3, we further mathematically formulate the unsupervised training strategy and generate the intuition of the correctness of the process.

3.1   AutoEmbedder Framework

The training process of AutoEmebdder includes a siamese network architecture. Formally the framework can be presented as,

$$\mathcal{S}(x, x^{'}) = ReLU(\|\mathcal{E}_\phi(x) - \mathcal{E}_\theta(x^{'})\|, \alpha) = \mathbb{R}^{+}_{\leq \alpha} \tag{1}$$

The ReLU function used in Eq. 1 is a thresholded ReLU function, such that,

$$ReLU(x, \alpha) = \begin{cases} x & \text{if } 0 \leq x < \alpha \\ \alpha & \text{if } x \geq \alpha \end{cases} \tag{2}$$

In Eq. 1, the $\mathcal{S}(\cdot, \cdot)$ represents a siamese network function, that receives two inputs. The framework contains two separate DNN networks (with the same initial weights) defined as $\mathcal{E}_\phi(\cdot)$ and $\mathcal{E}_\theta(\cdot)$ that map higher dimensional input to a lower dimension clusterable embedding vectors. The Euclidean distance of the embedding vector pairs is calculated and passed through the thresholded ReLU activation function derived in Eq. 2. The threshold value is a cluster margin of α. Considering $\mathcal{C}_p$ and $\mathcal{C}_q$ as two different cluster assignments, the pairwise constraint of the AutoEmbedder can be defined as,

$$\mathcal{Y} = \mathcal{P}_c(x, x^{'}) = \begin{cases} 0 & \text{if } x, x^{'} \in \mathcal{C}_p \\ \alpha & \text{if } x \in \mathcal{C}_p \text{ and } x^{'} \in \mathcal{C}_q \end{cases} \tag{3}$$

The overall architecture is trained using backpropagation method, where, the loss is calculated as,

$$L_p = \frac{1}{n} \sum_{i}^{n} \left\| \mathcal{Y}_i - \mathcal{S}(x_i, x^{'}_i) \right\| \tag{4}$$

Furthermore, in each training batch, selecting half data pair of similar clusters and half data pair with different cluster assignment improves the accuracy of the AutoEmbedder. For each data pair, the training of the siamese network is executed as follows,

$$S.train(x_i, x_i', \mathcal{Y}_i)$$
$$S.train(x_i', x_i, \mathcal{Y}_i) \tag{5}$$

Through the training process, each of the AutoEmbedder mimics similar embeddings for a given image. Whereas, for distinctive image pairs, the embedding vectors are formed at a minimum distance of α.

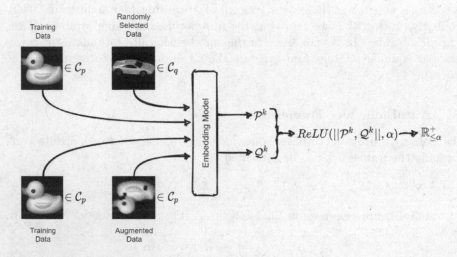

Fig. 1. The figure illustrates the basic intuition of the unsupervised strategy. For a given dataset, a randomly selected data pair is assumed relating to different clusters $(\mathcal{C}_p, \mathcal{C}_q)$. A data and the augmented version of the data are considered to refer to the same cluster $(\mathcal{C}_p)$.

3.2 Unsupervised Learning Strategy

To train the UAutoEmbedder in an unsupervised method, we can generate the pairwise constraints based on two assumptions,

1. A pair of data x and x' is chosen randomly, and it is assumed that $x \in \mathcal{C}_p$ and $x' \in \mathcal{C}_q$. These pairs are defined as cannot-link pairs.
2. A clean data x_i and the augmented instance of the data $Augment(x')$ are considered to belong to the same cluster $x, Augment(x') \in \mathcal{C}_p$. These pairs are expressed as must-link pairs.

The first assumption is dependent on the random selection of two pairs. A cannot-link criterion is implemented in AutoEmbedder to identify the dissimilar images. However, in the case of unsupervised learning, a dissimilarity criterion must be developed to instruct the model on image and class dissimilarity. Otherwise, the model would only be biased over similar class pairs. In a uniformly distributed dataset, if two images are randomly picked, it is validated that most of the image pairs would belong to different classes (discussed in Sect. 3.3). Due to randomness, it is also observed that such must-link pairs $x, x' \in C_p$ are also selected as cannot-link pairs, $x \in C_p, x' \in C_q$. However, the erroneous pair selection is less frequent than the correct pair selection (presented in Sect. 3.3). Therefore, the less frequent erroneous pairs are often neglected by the CDNN model.

The second assumption of generating must-link data pairs is related to image augmentation. Image augmentations can be applied using fundamental geometric transformations and distortions or by using DL approaches [31]. Often, DL approaches require training on similar data domain. Therefore, we avoid implementing DL based augmentation. Augmentation performs a vital objective in generating optimal clusterable embeddings. If augmentation produces a high diversity of data, while including the essential features, it may assist the UAutoEmbedder to converge to an optimal state. However, the selection of augmentation methods must be made based on the requirement of the dataset. In the Case study (Sect. 5.2), we conducted a thorough analysis exploiting the properties of augmentation considering the dataset specifications.

The fundamental algorithm of the AutoEmbedder architecture is generated using two parallel DCNNs $\mathcal{E}_\phi$ and $\mathcal{E}_\theta$ (does not share weights) with the same architecture. Instead, we implement the siamese network using DCNN with shared weights. The weight share causes a reduction of trainable parameters. Furthermore, instead of swapped pair training (Eq. 5), we apply random swaps of the training data pair. Excluding the swapped pair training causes the chance of overfitting on the inaccurate randomly selected cannot-link pairs. The pseudocode of the unsupervised training strategy is presented in Algorithm 1.

3.3 Mathematical Analysis

Let us consider a greedy CDNN function $\mathcal{E}_\phi(\cdot)$, and it is used as an embedder. The single embedder is used to build a siamese architecture $\mathcal{S}_\phi(\cdot, \cdot)$, that is,

$$\mathcal{S}_\phi(x, x') = ReLU(||\mathcal{E}_\phi(x) - \mathcal{E}_\phi(x')||, \alpha) = \mathbb{R}^+_{\leq \alpha} \tag{6}$$

The $\mathcal{S}_\phi(\cdot, \cdot)$ is further trained using backpropagation, for which, the loss is defined as,

$$L_p = \frac{1}{n} \sum_i^n \left\| \mathcal{Y}_i - S(x, x') \right\| \tag{7}$$

As the function $\mathcal{S}_\phi(\cdot, \cdot)$ reduces the loss value through backpropagation, the $\mathcal{E}_\phi(\cdot)$ must learn the higher-order features from the augmented data and generate a local cluster by combining the clean and the augmented data embeddings. The local clusters of similar characteristics are merged based on the higher-order feature discovery and the generalization of the CDNN architecture. However, to

---

Algorithm 1: UAutoEmbedder training pseudocode.

Input: Dataset D, UAutoEmbedder model with initial weights $\mathcal{E}_\phi$, Distance hyperparameter α, Training batch size B_s, Training epochs E_p

Output: Trained UAutoEmbedder

$\mathcal{S}_\phi(\cdot,\cdot) \leftarrow ReLU(||\mathcal{E}_\phi(\cdot) - \mathcal{E}_\phi(\cdot)||, \alpha)$
for $epoch \leftarrow 1$ **to** E_p **do**
   for $batch \leftarrow 0$ **to** $|D|/B_s - 1$ **do**
      $\mathcal{X}, \mathcal{X}', \mathcal{Y} \leftarrow \{\}, \{\}, \{\}$
      for $i \leftarrow 1$ **to** $B_s/2$ **do**
         $x \leftarrow \{D_{batch \times B_s + i}\}$
         $x' \leftarrow \{D_{\mathbb{N} \cap [1, |D|]}\}$
         if $\mathbb{N} \cap [0, 1] == 1$ **then**
            $x, x' \leftarrow x', x$
         $\mathcal{X} \leftarrow \mathcal{X} \cup x$
         $\mathcal{X}' \leftarrow \mathcal{X}' \cup x'$
         $\mathcal{Y} \leftarrow \mathcal{Y} \cup \alpha$
      for $i \leftarrow (B_s/2) + 1$ **to** B_s **do**
         $x \leftarrow \{D_{batch \times B_s + i}\}$
         $x' \leftarrow \{Augment(D_{batch \times B_s + i})\}$
         if $\mathbb{N} \cap [0, 1] == 1$ **then**
            $x, x' \leftarrow x', x$
         $\mathcal{X} \leftarrow \mathcal{X} \cup x$
         $\mathcal{X}' \leftarrow \mathcal{X}' \cup x'$
         $\mathcal{Y} \leftarrow \mathcal{Y} \cup 0$
      $S_\phi \leftarrow S_\phi.train(\mathcal{X}, \mathcal{X}', \mathcal{Y})$

---

further increase the probability of merging local clusters, the UAutoEmbedder is initialized with a pre-trained weight trained on any known dataset.

The ground truth $\mathcal{Y}$ of the Eq. 7 is initialized as,

$$\mathcal{Y} = \mathcal{P}_c(x, x') = \begin{cases} 0 & \text{if } x_i, Augment(x_i) \in \mathcal{C}_p \\ \alpha & \text{if } x_i \in \mathcal{C}_p \text{ and } x_j \in \mathcal{C}_q \end{cases} \tag{8}$$

However, the ground truth generated through the $\mathcal{Y}$ is erroneous due to the assumption that randomly selecting a data pair x_i and x_j will always belong to a different cluster. The training algorithm 1 describes that to train the function $\mathcal{S}_\phi(\cdot,\cdot)$, $B_s/2$ random pairs are always selected. As the function $\mathcal{S}_\phi(\cdot,\cdot)$ always converges to the smallest loss state, we can assume that if the probability of selecting incorrect pairs are always less than the correct pairs, the function $\mathcal{S}_\phi(\cdot,\cdot)$ can converge to an optimal state.

Let us consider a dataset D consisting of N_c classes where each class contains a uniform number of data N_p. The probability of selecting an erroneous pair is,

$$S_e = \frac{N_c \times P(N_p, 2)}{P(|D|)}$$
$$= \frac{N_c \times N_p \times (N_p - 1)}{|D| \times (|D| - 1)}$$
$$\approx \frac{N_c \times N_p^2}{|D|^2}$$
$$\approx \frac{N_c \times N_p^2}{(N_c \times N_p)^2}$$
$$\approx \frac{1}{N_c} \tag{9}$$
$$and, \quad S_e < \frac{1}{N_c} \quad [N_c > 1]$$

Therefore, for any dataset containing multiple classes, the value of selecting erroneous cannot-link pairs is always less than the correctly chosen cannot link pairs. Hence, it can be concluded that if the function $\mathcal{S}_\phi(\cdot, \cdot)$ converges to a minimal loss value, it can adequately separate cannot-link class pairs.

4 Experimental Setup

4.1 Evaluation Metrics

For evaluating, three clustering measures, Clustering Accuracy (ACC), Normalized Mutual Information (NMI), and Adjusted Rand Index (ARI) have been applied. All the metrics generate values in the range [0, 1]. A higher value of these metrics indicates a better clustering performance.

4.2 Datasets

Five datasets are used to evaluate the unsupervised AutoEembedder. Short descriptions of the datasets are given below:

MNIST: The dataset contains images of handwritten digits from 0 to 9. It contains 60, 000 handwritten digits. Each image is of shape 28×28 pixels in grayscale format.

Fashion-MNIST: The dataset contains images of different items of clothing and shoes with a total of ten categories. It contains 60, 000 grayscale images of 28×28 pixels.

CIFAR10: The dataset contains images of vehicles and animals, distributed to ten categories. It contains 50, 000 colorful images of 32×32 pixels.

COIL20: The dataset contains images of 20 different household objects. It has a total of 1, 440 grayscale images. For the experiment, we used the default dataset image shape of 128×128 pixels.

COIL100: The dataset is an extended version of the COIL20 dataset. It contains a total of 100 different classes. The dataset includes 7, 200 colorful images, which were of size 128 × 128 pixels, which is the default setup of the dataset.

4.3   Environmental Setup

The DCNN pipeline of the UAutoEmbedder architecture is generated using *TensorFlow* [1] and *Keras* [9]. As the UAutoEmbedder only generates the embeddings of each image, k-means clustering is performed to assign pseudo-labels of the embedding points. *scikit-learn* [26] implemented k-means algorithm is used in the evaluation. MobileNet [13] architecture is used as the DCNN baseline. A dense layer is attached to the last layer of the MobileNet [13] architecture to generate lower dimension embeddings. Twelve-dimensional embedding space is used for the overall evaluation of the UAutoEmbedder framework. *albumnetation* [2] is used to generate augmentation. The backpropagation is performed using Adam [19] optimizer with a learning rate of 0.001. A default batch size 128 is used with a limit of 400 epochs.

5   Result and Analysis

5.1   Comparison

Table 1 depicts the comparison of UAutoEmbedder with different unsupervised image clustering methods. The highest scores are marked in bold. The comparison demonstrates that most unsupervised methods produce a decent accuracy in the MNIST and Fashion-MNIST dataset. Yet, the UAutoEmbedder does not provide better efficiency in the MNIST dataset. Furthermore, in CIFAR10 dataset, most of the architectures show insufficient accuracy. The inefficiency on CIFAR10 dataset is reasonable. CIFAR10 contains environmental backgrounds, and therefore, it is difficult for unsupervised systems to generate the context of identifying correct features excluding the environmental properties. In contrast, the unsupervised methods produce sufficient accuracy in COIL20 and COIL100 datasets. The UAutoEmbedder generates better efficiency in most of the dataset, excluding the CIFAR10 and MNIST dataset. Although the inefficiency on CIFAR10 dataset is reasonable, the inability of producing better accuracy in MNIST dataset is illogical. The next section expresses a detailed logic of the uncertainties of the UAutoEmbdder.

Table 1. The table represents ACC, NMI, and ARI metric comparison of several unsupervised systems experimented on MNIST, Fashion-MNIST, CIFAR10, COIL20, and COIL100 dataset.

Model	MNIST			Fashion-MNIST			CIFAR10		
	ACC	NMI	ARI	ACC	NMI	ARI	ACC	NMI	ARI
JULE [35]	0.95 ± 0.02	0.97 ± 0.02	0.93 ± 0.03	0.53 ± 0.01	0.58 ± 0.08	0.39 ± 0.06	0.26 ± 0.05	0.32 ± 0.07	0.12 ± 0.01
VaDE [16]	0.94 ± 0.01	0.95 ± 0.02	0.88 ± 0.03	0.57 ± 0	0.59 ± 0.01	0.42 ± 0.03	0.31 ± 0.02	0.35 ± 0.04	0.27 ± 0.03
DEPICT [10]	0.92 ± 0.01	0.93 ± 0.01	0.87 ± 0.02	0.59 ± 0.02	0.63 ± 0.03	0.48 ± 0.03	$\mathbf{0.43 \pm 0.02}$	$\mathbf{0.48 \pm 0.03}$	$\mathbf{0.38 \pm 0.03}$
DBC [21]	0.92 ± 0.01	0.93 ± 0.03	0.91 ± 0.02	0.63 ± 0.03	0.65 ± 0.04	0.52 ± 0.03	0.34 ± 0.01	0.36 ± 0.02	0.22 ± 0.03
DEC-DA [12]	0.91 ± 0	0.92 ± 0.01	0.86 ± 0.01	0.56 ± 0.02	0.64 ± 0.03	0.49 ± 0.04	0.19 ± 0.05	0.24 ± 0.03	0.15 ± 0.03
SpectralNet [30]	0.92 ± 0.01	0.91 ± 0.01	0.95 ± 0.02	0.56 ± 0.01	0.63 ± 0.04	0.47 ± 0.04	0.22 ± 0.02	0.32 ± 0.04	0.18 ± 0.02
ClusterGAN [23]	0.91 ± 0.01	0.86 ± 0.06	0.89 ± 0.04	0.62 ± 0.02	0.65 ± 0.07	0.50 ± 0.08	0.41 ± 0.03	0.46 ± 0.04	0.30 ± 0.02
DAE Network [36]	0.93 ± 0.02	0.93 ± 0.04	0.91 ± 0.03	0.63 ± 0.01	0.61 ± 0.07	0.53 ± 0.04	0.42 ± 0.01	0.48 ± 0.04	0.31 ± 0.03
RTM [24]	0.94 ± 0.01	0.93 ± 0.03	0.91 ± 0.02	0.69 ± 0.04	0.68 ± 0.6	0.58 ± 0.02	0.30 ± 0.02	0.31 ± 0.08	0.23 ± 0.03
DDC [28]	$\mathbf{0.96 \pm 0.02}$	$\mathbf{0.97 \pm 0.01}$	$\mathbf{0.94 \pm 0.01}$	0.60 ± 0.03	0.67 ± 0.02	0.56 ± 0.02	0.36 ± 0.01	0.40 ± 0.01	0.30 ± 0.04
UAutoEmbedder	0.37 ± 0.01	0.33 ± 0	0.21 ± 0	$\mathbf{0.71 \pm 0.02}$	$\mathbf{0.74 \pm 0.01}$	$\mathbf{0.67 \pm 0.02}$	33.1 ± 1.9	0.36 ± 0.03	0.27 ± 0.05
Model	COIL20			COIL100					
	ACC	NMI	ARI	ACC	NMI	ARI			
JULE [35]	0.89 ± 0.02	0.96 ± 0.04	0.88 ± 0.02	0.79 ± 0.05	0.87 ± 0.03	0.73 ± 0.03			
VaDE [16]	0.88 ± 0.01	0.89 ± 0.03	0.79 ± 0.02	0.84 ± 0.01	0.89 ± 0.02	0.70 ± 0.01			
DEPICT [10]	0.69 ± 0.01	0.71 ± 0.03	0.61 ± 0.02	0.61 ± 0.03	0.69 ± 0.01	0.58 ± 0.01			
DBC [21]	0.77 ± 0.02	0.85 ± 0.03	0.63 ± 0.01	0.75 ± 0.01	0.80 ± 0.02	0.69 ± 0.01			
DEC-DA [12]	0.59 ± 0.02	0.72 ± 0.08	0.52 ± 0.04	0.58 ± 0.02	0.68 ± 0.03	0.50 ± 0.01			
SpectralNet [30]	0.70 ± 0.01	0.72 ± 0.07	0.63 ± 0.02	0.67 ± 0.02	0.70 ± 0.02	0.56 ± 0.03			
ClusterGAN [23]	0.78 ± 0.01	0.80 ± 0.02	0.70 ± 0.01	0.72 ± 0.01	0.75 ± 0.01	$0.68 \pm .01$			
DAE Network [36]	0.69 ± 0.02	0.72 ± 0.04	0.60 ± 0.04	0.65 ± 0.01	$0.70 \pm .02$	0.58 ± 0.01			
RTM [24]	0.82 ± 0.03	0.70 ± 0.09	0.68 ± 0.03	0.76 ± 0.03	0.81 ± 0.01	0.73 ± 0			
DDC [28]	0.58 ± 0.01	0.63 ± 0	0.51 ± 0.09	0.52 ± 0.02	0.58 ± 0.02	0.48 ± 0.01			
UAutoEmbedder	$\mathbf{0.96 \pm 0.01}$	$\mathbf{0.98 \pm 0.01}$	$\mathbf{0.93 \pm 0.02}$	$\mathbf{0.91 \pm 0.01}$	$\mathbf{0.97 \pm 0.01}$	$\mathbf{0.90 \pm 0.01}$			

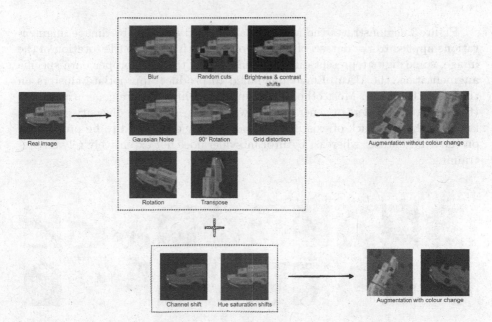

Fig. 2. The figure visualizes the various geometrical transformations applied to generate augmented images. Brightness and contrast shifts, noises are also included in the augmentation. An illustration of the colour change through augmentation is also visualized. Yet, the evaluation is conducted without any colour shifts.

5.2 Case Study

In Fig. 2, an example of the applied augmentation is illustrated. The experiment is conducted using the augmentations that do not alter the colour of the image. Therefore, no channel, hue, and saturation shifts have been conducted. Furthermore, online augmentation (e.g. random generation of augmentation for each batch) is applied to utilize the diversity of augmentation properly.

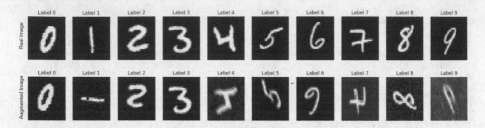

Fig. 3. The figure illustrates the MNIST dataset. The upper row visualizes the ten handwritten digits, whereas the lower row represents the augmented images of the respective handwritings. The augmented image of number 6 is similar to 9. Whereas, the augmented image of 7 is alike a reflected 4. This example demonstrates that the incorrect selection of augmentation techniques may generate inaccurate features.

Figure 3 demonstrates the MNIST dataset and the possible image augmentations applied to the dataset. It can be observed that due to the rotation of the image, some digits represent different senses. Due to the improper data-specific augmentation, the UAutoEmbedder failed to produce appropriate clusters on the MNIST dataset. Most other clustering algorithms require generative noises (through autoencoders or generative networks) to represent a diversity of similar images. The drawback of existing methods is they often need to be pre-trained on the same dataset whereas, UAutoEmbedder does not extensively rely on pre-training.

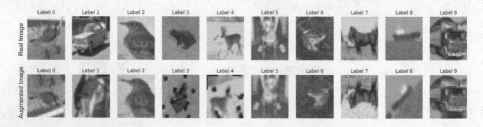

Fig. 4. The figure illustrates the CIFAR10 dataset. The upper row represents the clean images of the dataset of ten individual classes, whereas the lower row shows a possible augmentation for each of the corresponding images. Due to the different background images, the UAutoEmbedder can not correctly cluster the CIFAR10 dataset.

Figure 4 illustrates the CIFAR10 dataset along with the possible image augmentations. The UAutoEmbedder almost fails to cluster CIFAR10 dataset. However, nearly all of the unsupervised methods fail to generate a better score in CIFAR10 dataset. The reason of improper clustering depends on the environmental properties of the dataset. Due to the various environmental features, often the unsupervised strategies fails to identify appropriate features that are required to generate relevant clusters. The UAutoEmbedder mostly suffers from properly nullifying background features of an object.

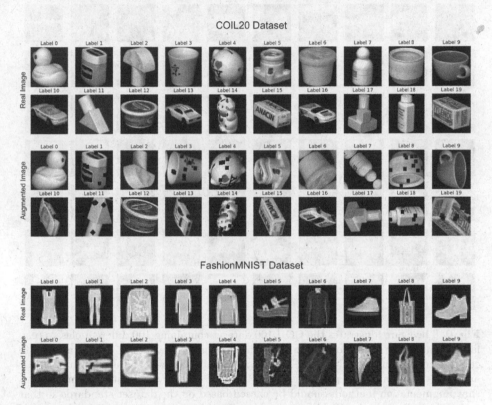

Fig. 5. The figure illustrates the COIL20 and Fashion-MNIST dataset. For each dataset image, a possible augmentation is presented in the corresponding row.

Using the explained augmentation, the UAutoEmbedder outperformed most of the unsupervised strategies in Fashion-MNIST, COIL20, and COIL100 dataset. Figure 5 illustrates the real and possible augmented images applied to COIL20 and Fashion-MNIST dataset. However, the criteria of augmentation had an essential influence on COIL100 dataset.

Fig. 6. The figure illustrates the COIL100 dataset containing 100 different classes (e.g. label 7 and 16 are similar objects with distinct colours). The dataset assigns a separate category depending on colour and shapes. As colour is considered as a feature, colour augmentation cannot be performed. This type of category separation demonstrates that augmentation methods should be chosen based on the dataset standards so that it does not alter the classifiable components of the image.

Figure 6 represents the 100 unique classes of the COIL100 dataset. The COIL100 distinguishes similar objects with different colours. Therefore similar objects with different colours are also considered in a different class. Due to this certain criteria, merging colour shifts with the applied augmentation resulted in a 5% reduction of the Autoemebedder accuracy comparing to the accuracy reported in Table 1.

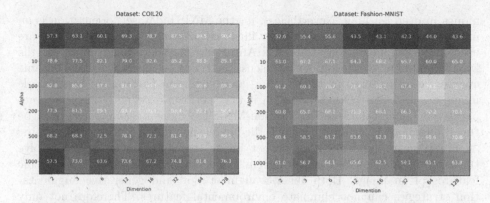

Fig. 7. The figure presents the impact on the accuracy w.r.t the hyperparameter α and the output dimension of the UAutoEmbedder. The analysis is carried on a subset of the Fashion-MNIST and COIL20 dataset. The grid search demonstrate that selecting $\alpha = [100, 200]$ often generates better results.

5.3 Sensitivity Analysis

The semi-supervised AutoEmbedder performed better with $\alpha = 100$ and output of three dimensions. However, the UAutoEmbedder is sensitive w.r.t to the hyperparameter α and the output dimensions. Figure 7 exhibits a grid search to investigate the dependency of the α and output dimension. The experiments are done in a smaller subset of the available dataset to reduce the time complexity of the grid search. For both of the datasets selecting $\alpha = [100, 200]$ generates better results. On the other hand, selecting the output dimension within $[12, 64]$ generates better accuracy. Therefore, choosing an output dimension within $[12, 64]$ and $\alpha = [100, 200]$ will result in generating a consistent result of the UAutoEmbedder.

6 Discussion

Considering the unsupervised image clustering of the UAutoEmbedder, it has numerous applications, including object image clustering, facial image clustering, image retrieval, and so on [33]. Although the UAutoEmbedder fails to cluster the MNIST dataset correctly, it has opportunities for object-based image clustering tasks. The selection of augmentation methods can turn UAutoEmbedder into a proper image clustering and similarity search method.

From the performance analysis, it can be considered that the selection of augmentation procedures largely depends on the property of the dataset. For instance, selecting color shifting for augmentation may result in UAutoEmbedder discriminating between colors. Objects in COIL100 have a dependency on colors. Consequently, if color augmentation is performed in COIL100, the UAutoEmbedder would perform inconsistently in classes depending on color. Horizontal/vertical flips and rotations are generic but powerful augmentation techniques in deep learning. Nevertheless, for the MNIST dataset, orientation is an important feature that is hindered by flips and rotates. As flips and rotates are used as an augmentation strategy, UAutoEmbedder performed poorly in the MNIST dataset. As a result of this discussion, it can be concluded that the same augmentation policies might not be a good choice to perform on very simple image datasets like MNIST although it may perform excellently in other comparatively difficult datasets like Fashion-MNIST. Augmentation policies must be tuned based on the criteria by which the model should differentiate image pairs.

Furthermore, implementing proper image segmentation [4] would resolve the issue of laboriously considering environmental features than class-specific features of an image. Both generic augmentation and generative augmentation strategies can not eliminate environmental features. Therefore not only UAutoEmbedder but also other models perform inadequately in the CIFAR10 dataset. As the UAutoEmbedder has been validated to perform better in all of the segmented image datasets (Fashion-MNIST, COIL20, COIL100), with segmentation, the UAutoEmbedder can be implemented further in real-world image clustering tasks.

7   Conclusion

The paper introduces UAutoEmbedder which resembles low-dimensional clusterable embeddings, generated from high-dimensional image data. The training process of the UAutoEmbedder is unsupervised and depends on pairwise constraints. The unsupervised training of the AutoEmbedder incorporates augmented image data to define must-link pairs, and randomly chooses data for cannot link pairs. Instead of performing any traditional clustering loss, the training is conducted using Euclidean distance loss, which is further minimized using backpropagation. Extensive experiments on various image datasets demonstrate the possibilities and challenges of the unsupervised training procedure. Moreover, a thorough analysis is conducted in search of optimal hyperparameters. The unsupervised framework achieves superior accuracy in COIL20, COIL100, and Fashion-MNIST datasets. We strongly believe that the overall contribution of the research work guides toward an innovative relationship between unsupervised learning and image data augmentation.

References

1. Abadi, M., et al.: Tensorflow: a system for large-scale machine learning. In: 12th {USENIX} Symposium on Operating Systems Design and Implementation ({OSDI} 16), pp. 265–283 (2016)
2. Buslaev, A., Iglovikov, V.I., Khvedchenya, E., Parinov, A., Druzhinin, M., Kalinin, A.A.: Albumentations: fast and flexible image augmentations. Information **11**(2), 125 (2020)
3. Chang, J., Wang, L., Meng, G., Xiang, S., Pan, C.: Deep adaptive image clustering. In: Proceedings of the IEEE International Conference on Computer Vision, pp. 5879–5887 (2017)

4. Chen, M., Artières, T., Denoyer, L.: Unsupervised object segmentation by redraw-ing. In: Advances in Neural Information Processing Systems, pp. 12726–12737 (2019)
5. Chen, X., Duan, Y., Houthooft, R., Schulman, J., Sutskever, I., Abbeel, P.: Infogan: Interpretable representation learning by information maximizing generative adver-sarial nets. In: Advances in Neural Information Processing Systems, pp. 2172–2180 (2016)
6. Dai, Z., Yang, Z., Yang, F., Cohen, W.W., Salakhutdinov, R.R.; Good semi-supervised learning that requires a bad GAN. In: Advances in Neural Information Processing Systems, pp. 6510–6520 (2017)
7. Deng, J., Dong, W., Socher, R., Li, L.J., Li, K., Fei-Fei, L.: Imagenet: a large-scale hierarchical image database. In: 2009 IEEE Conference on Computer Vision and Pattern Recognition, pp. 248–255. IEEE (2009)
8. Ester, M., Kriegel, H.P., Sander, J., Xu, X., et al.: A density-based algorithm for discovering clusters in large spatial databases with noise. In: KDD, vol. 96, pp. 226–231 (1996)
9. Géron, A.: Hands-on machine learning with Scikit-Learn, Keras, and TensorFlow: Concepts, tools, and techniques to build intelligent systems. O'Reilly Media (2019)
10. Ghasedi Dizaji, K., Herandi, A., Deng, C., Cai, W., Huang, H.: Deep clustering via joint convolutional autoencoder embedding and relative entropy minimization. In: Proceedings of the IEEE International Conference on Computer Vision, pp. 5736–5745 (2017)
11. Goodfellow, I., et al.: Generative adversarial nets. In: Advances in Neural Infor-mation Processing Systems, pp. 2672–2680 (2014)
12. Guo, X., Zhu, E., Liu, X., Yin, J.: Deep embedded clustering with data augmen-tation. In: Asian Conference on Machine Learning, pp. 550–565 (2018)
13. Howard, A.G., et al.: Mobilenets: efficient convolutional neural networks for mobile vision applications. arXiv preprint arXiv:1704.04861 (2017)
14. Hsu, C.C., Lin, C.W.: CNN-based joint clustering and representation learning with feature drift compensation for large-scale image data. IEEE Trans. Multimedia **20**(2), 421–429 (2017)
15. Hu, W., Miyato, T., Tokui, S., Matsumoto, E., Sugiyama, M.: Learning dis-crete representations via information maximizing self-augmented training. arXiv preprint arXiv:1702.08720 (2017)
16. Jiang, Z., Zheng, Y., Tan, H., Tang, B., Zhou, H.: Variational deep embed-ding: an unsupervised and generative approach to clustering. arXiv preprint arXiv:1611.05148 (2016)
17. Jin, X., Han, J.: K-means clustering. In: Encyclopedia of Machine Learning and Data Mining, pp. 695–697. Springer, Berlin, Heidelberg (2017). https://doi.org/10.1007/978-3-642-29807-3
18. Johnson, S.C.: Hierarchical clustering schemes. Psychometrika **32**(3), 241–254 (1967)
19. Kingma, D.P., Ba, J.: Adam: a method for stochastic optimization. arXiv preprint arXiv:1412.6980 (2014)
20. Kingma, D.P., Welling, M.: Auto-encoding variational bayes. arXiv preprint arXiv:1312.6114 (2013)
21. Li, F., Qiao, H., Zhang, B.: Discriminatively boosted image clustering with fully convolutional auto-encoders. Pattern Recogn. **83**, 161–173 (2018)
22. Mridha, M.F., Ohi, A.Q., Monowar, M.M., Hamid, M.A., Islam, M.R., Watanobe, Y.: U-vectors: Generating clusterable speaker embedding from unlabeled data. Appl. Sci. **11**(21), 10079 (2021)

23. Mukherjee, S., Asnani, H., Lin, E., Kannan, S.: ClusterGAN: latent space clustering in generative adversarial networks. In: Proceedings of the AAAI Conference on Artificial Intelligence, vol. 33, pp. 4610–4617 (2019)

24. Nina, O., Moody, J., Milligan, C.: A decoder-free approach for unsupervised clustering and manifold learning with random triplet mining. In: Proceedings of the IEEE International Conference on Computer Vision Workshops (2019)

25. Ohi, A.Q., Mridha, M., Safir, F.B., Hamid, M.A., Monowar, M.M.: Autoembedder: a semi-supervised DNN embedding system for clustering. Knowledge-Based Systems, p. 106190 (2020)

26. Pedregosa, F., et al.: Scikit-learn: machine learning in python. J. Mach. Learn. Res. **12**, 2825–2830 (2011)

27. Ren, Y., Hu, K., Dai, X., Pan, L., Hoi, S.C., Xu, Z.: Semi-supervised deep embedded clustering. Neurocomputing **325**, 121–130 (2019)

28. Ren, Y., Wang, N., Li, M., Xu, Z.: Deep density-based image clustering. Knowledge-Based Systems, p. 105841 (2020)

29. Schroff, F., Kalenichenko, D., Philbin, J.: Facenet: a unified embedding for face recognition and clustering. In: Proceedings of the IEEE Conference on Computer Vision and Pattern Recognition, pp. 815–823 (2015)

30. Shaham, U., Stanton, K., Li, H., Nadler, B., Basri, R., Kluger, Y.: Spectralnet: spectral clustering using deep neural networks. arXiv preprint arXiv:1801.01587 (2018)

31. Shorten, C., Khoshgoftaar, T.M.: A survey on image data augmentation for deep learning. J. Big Data **6**(1), 60 (2019)

32. Springenberg, J.T.: Unsupervised and semi-supervised learning with categorical generative adversarial networks. arXiv preprint arXiv:1511.06390 (2015)

33. Wazarkar, S., Keshavamurthy, B.N.: A survey on image data analysis through clustering techniques for real world applications. J. Vis. Commun. Image Represent. **55**, 596–626 (2018)

34. Xie, J., Girshick, R., Farhadi, A.: Unsupervised deep embedding for clustering analysis. In: International Conference on Machine Learning, pp. 478–487 (2016)

35. Yang, J., Parikh, D., Batra, D.: Joint unsupervised learning of deep representations and image clusters. In: Proceedings of the IEEE Conference on Computer Vision and Pattern Recognition, pp. 5147–5156 (2016)

36. Yang, X., Deng, C., Zheng, F., Yan, J., Liu, W.: Deep spectral clustering using dual autoencoder network. In: Proceedings of the IEEE Conference on Computer Vision and Pattern Recognition, pp. 4066–4075 (2019)

Author Index

Printed in the United States
by Baker & Taylor Publisher Services

Printed in the United States
by Baker & Taylor Publisher Services